WATER TREATMENT

Susan E. Kegley, Doug Landfear, David Jenkins, Kome Shomglin

Student Manual

Water Treatment

HOW CAN WE MAKE OUR WATER SAFE TO DRINK?

Susan E. Kegley
Pesticide Action Network North America, San Francisco, CA

Doug Landfear
University of California, Berkeley, CA

David Jenkins
UNIVERSITY OF CALIFORNIA, BERKELEY, CA

Kome Shomglin
University of California, Berkeley, CA

W. W. NORTON & COMPANY
NEW YORK LONDON

This ChemConnections module has been developed under the direction of the ChemLinks Coalition, headed by Beloit College, and the ModularChem Consortium, headed by the University of California at Berkeley. This Material is based upon work supported by the National Science Foundation grants No. DUE-9455918 and DUE-9455924. Any opinions, findings, and conclusions or recommendations expressed in this material are those of the authors and do not necessarily reflect the views of the National Science Foundation, Beloit College or the Regents of the University of California.

For more information on the ChemConnections modules, please visit: http://www.wwnorton.com/college/Chemistry/chem_home.htm

This project was supported, in part, by
the Division of Undergraduate
Education of the
National Science Foundation.
Opinions expressed are those of the
authors and not necessarily those of
the Foundation

ISBN 0-393-92646-X

Printed in the United State of America

7 8 9 0

Acknowledgments

Many individuals have contributed valuable advice and ideas. The authors would like to thank Barbara Heaney, Eileen Lewis, Beth Abdella, Paul Jasien, Geoff Marcy, Angy Stacy, Marco Molinaro, Susan Walden, Edward Yeh, Steve Lawenda, Jack Bell, and Ron Rusay for their helpful input. Experimental development for some of the laboratory work was done by Benjamin Gross. The cover and icon art was done by Leigh Anne McConnaughey. Programming for the Equilibrium Simulator was done by Cora Estrada. Animations were created by Marco Molinaro, Susan Walden, Leigh Anne McConnaughey, and Joel Russell. We would particularly like to thank the students and teaching assistants in the Spring 1998 Chemistry 1A class at UC Berkeley for their help in testing the module.

HOW TO USE THIS GUIDEBOOK

This guidebook describes a real-world problem that you will explore for several weeks of your first-year college chemistry course. During these weeks, you will be challenged to develop an effective water treatment plan for a contaminated water supply. As you consider various options for your plan, you will learn about several key concepts in chemistry, including stoichiometry, the general principles of equilibrium, and the specifics of solubility and acid-base equilibria. In addition, you will gain important skills in developing and testing models, and in solving problems with multiple constraints and multiple solutions. The prerequisites for this module include using stoichiometry and writing chemical reactions and being familiar with introductory concepts of covalent and ionic bonding, Lewis dot structures, predicting molecular shapes using VSEPR, periodic trends, and basic reaction energetics (DH). Algebra is also required. For classes in which thermodynamics was covered previously, your instructor may choose to do Explorations 3D, 4G, and 7E, which illustrate the connection of thermodynamics to equilibrium.

This guidebook includes inquiry-based activities for the classroom, the laboratory, the computer laboratory, and homework. The curriculum requires active participation from you, not only in the laboratories but also in the classroom. In each of these settings, you will be recording observations from demonstrations, doing experiments, solving problems, discussing results, and constructing models to explain your observations and analyses.

Organization of the Module

The Module begins with a question about designing a water treatment plan: **How can we purify our water?** This Module Question provides a context for understanding and exploring chemistry concepts involving solution chemistry and equilibrium. Carefully chosen questions are used throughout the Module as springboards into guided inquiry and exploration of chemistry concepts.

Sessions. The Module is divided into 9 Sessions, each beginning with a Session Question. Each Session focuses on one aspect of the Module Question and guides you to consider issues that you need to understand to respond to the Module Question.

Explorations. The Session Question is examined in classroom, laboratory, and computer Explorations. Each Exploration begins with a question that considers the Session Question in more depth and in different ways. Your instructor will choose which Explorations to use depending on her/his goals and the needs of your class. There is no need to complete all Explorations.

Each Exploration is divided into five sections as follows:

1. **Creating the Context.** This section states the goal of the Exploration and frames the Exploration Question. The text discusses why the question is important, makes links to information discussed previously, gives background information, and describes what you can do to find out more.

2. **Preparing for Inquiry.** This section provides background information that you should know before you attempt the questions in the Developing Ideas section. Be sure you have read these sections thoroughly before coming to class.

3. **Developing Ideas.** This special section consists of some activity that probes deeper into understanding the concepts and leads you through a series of questions that help you think about the important concepts necessary to formulate a response to the Session and Exploration Questions. Answering these questions successfully will usually require mastery of the material in prior Explorations and a thorough reading of the Exploration text. Activities of the following types will be used as a centerpiece for this guided inquiry: 1) watching a demonstration; 2) watching a video or computer simulation; 3) working in groups to evaluate a data set for trends and significance; 4) engaging in a class discussion; 5) experimenting in the laboratory; or 6) using interactive software. Each of these types of activities is discussed below, along with the icons used to designate each.

4. **Working with the Ideas.** A set of questions is provided to help you explore further the observations and data that you have collected and to give you practice in problem-solving. Your instructor will choose which of these questions to use for interactive discussion, small group work, or homework.

5. **Making the Link.** In this section, you will reflect upon what you needed to know to respond to the Exploration Question and how your response has helped you provide a more complete response to the Session Question.

Making the Link. Each Session ends with a section called Making the Link. This section provides you with a list of chemistry concepts that you have learned and a list of problem-solving skills that are appropriate to put on a resume. In addition, a worked example in another context is often provided, using the concepts that you have learned. In this section you can Check Your Progress and think about how what you've learned has helped you understand the Module Question.

Culminating Project. The Module will end with a project in which you will be asked to integrate everything that you have learned to provide a detailed response to the Module Question.

Icons Used in the Guidebook

You will encounter several icons to indicate the type of activity that you will do in Gathering Information for each Exploration that you do. These icons are shown below, along with a brief explanation of what they signify.

Demonstration

Demonstrations will be performed by your instructor or shown on a video. It is important that you understand the purpose of the experiment, make detailed observations, and attempt to explain these observations.

Group work

Group discussion and class discussion will help you formulate responses to questions and give you different perspectives.

Laboratory experiment

In a laboratory exploration, you will design and carry out experiments in order to generate a response to a specific question. Doing these experiments will require that you use chemistry principles and thinking skills that are related to your work in the classroom.

Computer experiment

There are also assignments in which you will use a computer program that shows an animation of a chemical process. These programs are designed to help you understand ionic and molecular interactions and delve deeper into understanding the concepts of equilibrium.

Goals

The purpose of setting the module into the context of water treatment is to achieve the following goals:

1. To help you gain a better understanding of chemical concepts of solution chemistry and equilibrium by encouraging you to apply these concepts to water treatment.

2. To help you learn about the relationship between water quality and concentrations of dissolved species such as calcium, magnesium, iron, fluoride, pH, and total dissolved solids.

3. To help you learn the laboratory techniques used to measure and remove contaminants from water.

4. To help you gain a better understanding of what scientists do by designing solutions for complex problems with multiple constraints, such as making a water supply safe to drink.

Learning in a context. You may wonder about the necessity of providing a context such as water treatment. Perhaps you feel that you could just learn the concepts and not bother learning all the extra information about the context. Research in cognitive science has shown that it is worth the effort because we retain knowledge better when appropriate background material is available. As an example, try the following experiment. Read the following paragraph once, close the book, and write down as much as you remember.

> The procedure is not difficult. First, bring 1 liter of water to a state where it has partially undergone a phase transition in which the vapor pressure of the steam that is formed is equal to the pressure of the atmosphere. Then add 1.0 g of the mixture of chemicals known as camillea thea. The important ingredient in this mixture is 3,7-dihydro-1,3,7-trimethyl-1H-purine-2,6-dione. Allow the mixture to stir for 5 minutes. Finally, filter the undissolved solids and collect the liquid.

If you are unfamiliar with the technical language, it is difficult to recall much of these instructions. However, if you are told that the passage is about making tea, suddenly you can figure out much of the new vocabulary and enhance your retention of the instructions. The context has helped you use background knowledge to aid you in comprehending the passage. We have built a context surrounding the chemistry concepts and the problem-solving skills that we hope you will learn in this module. We hope that our use of this context will help you make links to your background knowledge and to experiences from your everyday life. We believe that making these connections will aid your comprehension and enhance your retention.

Transforming knowledge. As part of this module, you will also be encouraged to try to find solutions to problems that cannot be solved by simply "telling" knowledge. At this stage in your academic careers, it is important that you make the transition from simply "telling" facts that you have memorized to "transforming" your knowledge to solve new problems. This is what scientists and professionals such as doctors do. As an example, consider the type of problem that a medical doctor must solve.

A patient who is quite ill is examined by Dr. Rosario. After examining the patient the doctor realizes that he has never encountered this set of symptoms before, or at least he doesn't remember having encountered them. Imagine Dr. Rosario giving one of the following responses to the patient:

"Sorry, I can't help you. I can't find the answer in my textbook."

"Sorry, I can't help you. There is something wrong with your blood chemistry, but I can't remember what I learned in first-year chemistry."

"Sorry, I can't help you. I wasn't told about symptoms like yours in medical school."

Clearly, it is not possible for doctors or scientists to be familiar with every case or be able to remember all the necessary information. Therefore, it is critical that they know how to approach complex problems that do not present an immediate solution.

In this module, we hope to give you experience with approaches used by scientists to solve problems, especially those for which an answer is not immediately obvious and for which there may be multiple solutions. You will find that real-life problem solving is an iterative process. When you do not know at first how to solve a problem, you start by exploring your best ideas. If your initial ideas do not lead you toward solving the problem, you may have to back track, rethink your ideas, look up more information, and then try something else. This process of generating then refining your ideas allows you to define the problem more clearly. Eventually, you may reach an acceptable solution to the problem. As you generate and refine ideas about the issues and concepts in this module, your thinking will become more sophisticated, and your understanding will become deeper.

CONTENTS

SESSION 5: LE CHATELIER'S PRINCIPLE   99

How can we remove contaminants from a water supply?

SESSION 6: REMEDIATION   113

What procedures can you design to remove contaminants from a water supply?

SESSION 7: ACIDS AND BASES I   140

What are acids and bases?

SESSION 8: ACIDS AND BASES II   164

What is the role of acids and bases in water treatment?

SESSION 9: PROJECTS

What are your results?

APPENDIX: OPTIONAL EXPLORATION 3D

Introduction

How can we make our water safe to drink?

Exploration 1A: The Storyline

All living organisms require liquid water for survival. Throughout history, great civilizations have grown and thrived by developing natural supplies of water for drinking and irrigation. The same civilizations have declined and even disappeared when water supplies dried up or when irrigated fields became unproductive. It is for these reasons that a safe drinking water supply is a high priority for any society.

In this module, you will learn what is required to ensure a pure water supply for a community. You must first know how to determine whether toxic or harmful contaminants are in a water supply and learn something about their chemistry and how to quantify them. You will then explore the chemistry of some common water treatment processes and design methods for the removal of contaminants from a water supply. Your laboratory experience in the module will allow you to hone your problem-solving abilities as you design and implement your own plan for purifying a contaminated water supply.

BACKGROUND READING

Water Sources

As the population of the world continues to grow, water is an increasingly political issue. Nations that share water supplies must work out plans for all to have access to enough water for domestic and agricultural use. It may seem strange that the availability of water is a problem on a planet like ours, whose surface is about 70% covered by water. The problem is that most of this water is in the oceans (see Table 1-1), and ocean water is too salty to drink or use for irrigation. Other water supplies, called *freshwater*, might not be too salty, but they might contain high concentrations of other components such as pathogenic bacteria, heavy metals like cadmium or copper, or toxic organic compounds like petroleum products or dry cleaning fluid. In order to make a water supply safe, we need to know what substances are present and how to remove any undesirable contaminants.

In most instances, cities or towns have little flexibility in choosing their water sources. The closest **potable** (drinkable) water source is usually the cheapest and most convenient and might be:

1. Terrestrial **surface waters**, such as a river, stream, or lake. Such waters are major sources of water for human use.

2. **Groundwater**, or water trapped underground in layers of porous rock called **aquifers**. Such water is usually tapped by wells and is replenished by rainfall that percolates through the ground. The process of percolation filters and purifies the water and is one of the reasons that ground water is usually less polluted than surface water.

3.   A reservoir, which may store either surface water or ground water.

Table 1-1   Distribution of Water on the Earth's Surface

Source	Volume (thousands of km^3)	Percent of Total Water
Oceans	1,320,000	97.3
Icecaps and Glaciers	29,200	2.14
Groundwater	8,350	0.61
Freshwater Lakes	125	0.009
Saline Lakes and Inland Seas	104	0.008
Soil Moisture	67	0.005
Atmospheric Water	13	0.001
Rivers	1.25	0.0001

The Safe Drinking Water Act

To protect public health and ensure safe water supplies, Congress passed the Safe Drinking Water Act in 1974. Amended in 1986 and 1996, the law requires the Environmental Protection Agency (EPA) administrator to set standards for contaminant levels such that the concentrations of these species do not threaten public health and safety. The EPA regulates two sets of contaminants in water:

1.   **Primary contaminants:** those substances that may cause serious health problems above certain concentrations.

2.   **Secondary contaminants:** substances that are not toxic but detract from water quality, affecting such characteristics as appearance, smell, and taste.

The EPA sets goals for the **maximum contaminant levels (MCLs)** for both sets of contaminants and requires water treatment facilities to monitor concentrations of these substances on a regular basis. If treated water does not meet the minimum standards most of the time, water treatment engineers usually begin to re-evaluate the purity of their source water or find alternative treatments.

Concentration Units and Conversions

The best way to measure the amount of a substance dissolved in water is to report the amount of the substance (the **solute**) per unit volume of solvent (water, in this case). This value is known as the **concentration** of the substance. Several conventional units are used by chemists to express solute concentrations in a solvent. In many applications, the units used are **molarity (mol/L, or M)**, the number of moles of a substance in one liter of solution. Molar concentrations are represented by writing the species in square brackets. Thus, [Cl$^-$] is shorthand for saying "the concentration of chloride ion in moles per liter."

Often, however, the concentrations of dissolved substances in water supplies are given as the mass of the substance per liter of solution. Convenient units used for water quality parameters are **milligrams of solute per liter of solution (mg/L)**. Since 1 mg/kg = 1 mg/10^6 mg, we can say that 1 mg/kg = 1 **part per million (ppm)**. In dilute aqueous solutions, the units of mg/L can reasonably be equated with milligrams per kilogram (mg/kg), since one liter of water weighs one kilogram at room temperature. Thus, mg/L and ppm are often used interchangeably for dilute aqueous solutions. For smaller quantities of solutes, units of **parts per billion (ppb)** are used, where one ppb is equal to 1 μg/kg or 1 μg/L in dilute aqueous solution.

To convert from a concentration in g/L to a concentration in mol/L, it is necessary to divide the quantity of the substance in g/L (or mg/L) by the molecular weight or formula weight in g/mol (or mg/mmol). For example, suppose a solution contains 2.35 g NaCl/L. To express the concentration of NaCl in mol/L, we calculate the molecular weight of NaCl (58.5 g/mol) by adding the atomic weights of Na (23.0 g/mol) and Cl (35.5 g/mol). We then divide the solution concentration in g/L by the molecular weight in g/mol to obtain the solution concentration in mol/L.

$$\frac{2.35 \text{ g NaCl}}{L} \times \frac{1 \text{ mol}}{58.5 \text{ g NaCl}} = \frac{0.040 \text{ mol NaCl}}{L} = 0.040 \text{ M}$$

To convert from a concentration in mol/L to a concentration in g/L, simply reverse this procedure by multiplying the molar concentration by the molecular weight of the substance.

Often we wish to know the concentration of *only one* of the ions of a compound in a solution in g/L or mg/L. Using the example above, consider the situation in which you would like to know the concentration of *chloride* in the solution. To determine this value, we multiply the concentration of NaCl by the ratio of the molecular weights of chloride to sodium chloride.

$$\frac{2.35 \text{ g NaCl}}{L} \times \frac{35.45 \text{ g Cl}^-}{58.5 \text{ g NaCl}} = \frac{1.42 \text{ g Cl}^-}{L}$$

Developing Ideas

The first step in creating a usable water supply from a potential water source is to define the desirable characteristics of drinkable water. Working with your classmates:

1. List as many of these characteristics as you can. Discuss your list with the whole class.

2. With this list in mind, generate a step-by-step plan that would enable a community to transform an existing water source into a usable water supply. For now, you will need to work in generalities. You will enhance your plan as you learn more about the chemistry of water throughout the module.

Working with the Ideas

The following problems will help you understand the issues in more depth.

3. What does water purity mean to you? Is the best water supply composed of 100% H_2O? Explain.

4. What is the water source for your community?

5. What conditions might make a water source unusable as a drinking water supply, and how would you determine if those conditions prevail?

6. Go to the Environmental Protection Agency's web site at http://www.epa.gov/waterscience/drinking/standards/, download the pdf file with the table of water quality standards, and look at the Maximum Contaminant Levels (MCLs) for fluoride, iron, pH, and total dissolved solids. You will need to look in both the *Drinking Water Standards and Health Advisories* "Inorganics" list and the *Secondary Drinking Water Regulations* list for these values. Note: Fluoride appears on both lists.

7. If it were your job to set the MCL for a particular contaminant, what factors would you need to take into account to determine the amount of contaminant that is "too much"?

8. Even small quantities of heavy metals in water can pose a serious health risk. The MCLs for some of the heavy metals are as follows: lead, 15 ppb; arsenic, 0.01 ppm; mercury, 2 μg/L; cadmium, 5 ppb. Determine the MCLs of these metals in mol/L, assuming a dilute aqueous solution.

9. The MCL for cyanide (CN^-) is 0.2 mg/L and for nitrate (NO_3^-) is 10 mg/L. Convert these MCLs from mg/L to mol/L.

10. A drinking water supply was found to contain 0.01 M calcium and 0.002 M magnesium. What are the concentrations of these metals in mg/L (ppm)?

11. How might an understanding of chemistry be useful in purifying a water supply? Look at the table of contents for this module. Identify the chemical concepts that you need to carry out the step-by-step plan you generated.

Looking Ahead

Several Exploration Questions can guide you to explore various aspects of the Module Question in more depth. These will be used at the discretion of your instructor.

- **Exploration 1B:** What substances do you typically find in natural waters?
- **Exploration 1C:** How do dissolved substances get into a water supply?
- **Exploration 1D:** How can we obtain a quantitative profile of the ionic constituents of a water supply?

Exploration 1B

What substances do you typically find in natural waters?

Creating the Context

Why is this an important question?

The first step toward purifying a water supply is to find out what substances it contains. Does the water look cloudy? Does it smell? Are there constituents in the water that are invisible to our five senses? Are there birds, fish, and animals in or around the water source that might contribute their own wastes? Are there human activities that might add pollutants? If you are going to use a water source for a drinking water supply, you need to know what substances you might expect to find in natural waters, so you can begin to devise a plan to purify the water. Thus, the goal of this Exploration is to answer the question: *What substances do you typically find in natural waters?*

Developing Ideas

Your instructor will provide several water samples for you to observe.

1. Working with your classmates, use these samples to create a list of substances you would expect to find in natural waters. Classify them as natural or anthropogenic (from human activities).

2. Discuss your ideas with the class as a whole.

Working with the Ideas

The following problems will help you understand the issues in more depth.

3. What problems do animals, birds, and fish pose for a water supply?

4. What contaminants that were not present in the source water might be in your water by the time it comes out of the faucet?

5. Many constituents in water arise from human activities. List as many of these water pollutants as you can, along with the activities that cause or permit these substances to make their way into a water supply.

Exploration 1C

How do dissolved substances get into a water supply?

Creating the Context

Why is this an important question?

In Exploration 1B, we saw that a natural water supply might have many constituents. How do they get there? If we know the answer to this question, we can begin to think about ways to reduce the concentration of a substance in a water supply by preventing it from getting there in the first place. While this may be a workable treatment plan for some water supplies, we'll see that it won't necessarily work for all of them. In this Exploration, we will answer the question, *How do dissolved substances get into a water supply?*

Preparing for Inquiry

BACKGROUND READING

The Hydrologic Cycle

Let's begin by looking at the ways in which water moves through the environment. Water can exist as a solid in ice and snow; as a liquid in surface water, groundwater, and the oceans; and as a gas in the atmosphere. Water cycles through the environment by a variety of processes collectively known as the **hydrologic cycle**: evaporation, sublimation, evapotranspiration, condensation, and precipitation (see Figure 1-1).

Evaporation is the transformation of a liquid into a gas and requires energy from the sun or another source of heat. *Sublimation* is the transformation of a solid into a gas. Both evaporation and sublimation vaporize water from icecaps, snow fields, oceans, lakes, reservoirs, and rivers.

Evaporation also results from the metabolism of plants and from fuel combustion. Water is taken up by the root system of plants and eventually evaporated to the atmosphere through the leaves in a process called *evapotranspiration*. The combustion of fuels such as wood, oil, coal, gasoline, or natural gas also adds water vapor to the atmosphere.

$$\text{fuel} \ + \ \underset{\text{oxygen}}{O_2} \ \longrightarrow \ \underset{\substack{\text{carbon} \\ \text{dioxide}}}{CO_2} \ + \ \underset{\text{water}}{H_2O} \ + \ \underset{\substack{\text{nitrogen} \\ \text{oxides}}}{NO_x}$$

Condensation takes place in the atmosphere, as water vapor condenses on tiny particles to form water droplets that can fall to the earth's surface as precipitation in the form of rain, snow, hail, and sleet. *Precipitation* falls to the oceans and the land surface, from which evaporation takes place again to continue the cycle.

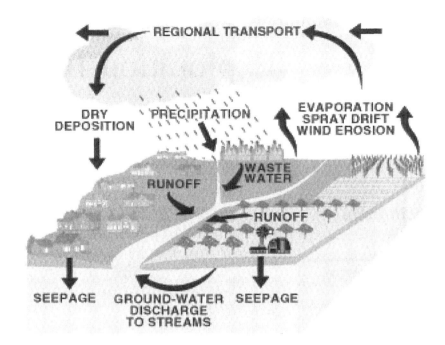

Figure 1-1:   The hydrologic cycle shows how water moves through the environment, evaporating from oceans, streams, land, and plants, then falling to earth as precipitation, where it travels through runoff back to surface waters.

Much of the water that falls onto the land surface forms **runoff,** or water that flows over the surface, eventually reaching streams, rivers, or lakes, where it is stored before evaporating or returning to the oceans. Some of the water that falls to the earth as precipitation passes through the soil to become part of the groundwater that is trapped in aquifers. Most aquifers are located within 800 meters of the earth's surface. The point at which groundwater meets the relatively dry soil above it is called the **water table**.

Effects of Runoff on Water Quality

When water falls on the earth's surface, it contacts rocks, soil, and organic matter. Rocks are made of minerals, which are composed of a variety of ionic inorganic compounds. As precipitation falls on rocks and soils, small amounts of these ions dissolve in the water. Runoff then carries these dissolved ionic substances into streams, rivers, lakes, and eventually the ocean. Table 1-2 gives the composition of some common minerals and can be used to predict the types of dissolved substances typically found in natural waters.

Table 1-2   Composition of Common Minerals

Mineral	Chemical Composition
apatite	$Ca_{10}(PO_4)_6(OH)_2$
biotite	$KMg_3[AlSi_3O_{10}](OH)_2$
calcite, aragonite (limestone)	$CaCO_3$
dolomite	$CaMg(CO_3)_2$
fayalite	Fe_2SiO_4
forsterite	Mg_2SiO_4
gypsum	$CaSO_4 \cdot 2H_2O$
halite	$NaCl$
hematite	Fe_2O_3
hornblende	$Ca_3Mg_5[Al_2Si_6O_{22}](OH)_2$
magnesite	$MgCO_3$
olivine	$(Fe,Mg)SiO_4$
orthoclase feldspars	$KAlSi_3O_8$
plagioclase feldspars	$NaAlSi_3O_8$, albite; $CaAl_2Si_2O_8$, anorthite
pyrite	FeS_2
quartz	SiO_2

Effects of Atmospheric Constituents on Water Quality

Many rocks and minerals are not very soluble in pure water; however, their solubility is often increased in the presence of acidic substances. A natural source of acids is the atmosphere. Let's take a closer look at the atmospheric gases and their effects on the concentration of dissolved ionic compounds in a water supply.

The major atmospheric gases, N_2 and O_2, are both sparingly soluble in water, 17.5 and 39.3 mg/L, respectively, at 25°C and a partial pressure of 1 atm. In contrast, some of the minor constituents of the atmosphere such as CO_2 and SO_2 are quite soluble, and even though their concentration in the atmosphere is much less than that of the major gases, their saturation concentrations in water are much higher, 1,450 and 94,100 mg/L, respectively, at 25°C.

The dissolution of CO_2 and SO_2 in water results in the formation of carbonic acid (H_2CO_3), sulfurous acid (H_2SO_3), and sulfuric acid (H_2SO_4). We will learn more about these acids in a later section, but for now, what you need to know is that acidic precipitation dissolves minerals much more readily than neutral precipitation and results in higher concentrations of dissolved ionic substances in runoff. In the natural hydrologic cycle, water evaporating from the oceans can be deposited as precipitation on the ocean and on land masses. During precipitation events, water passes through and equilibrates with gases present in the earth's lower atmosphere. Effects on water quality can vary because the earth's atmospheric composition varies, especially in the levels of minor constituents such as CO_2, SO_x ($x = 2, 3$), and NO_x ($x = 1, 2$), gases produced by combustion and associated with urban-industrial air pollution.

The composition of precipitation is also influenced by other atmospheric contaminants. Precipitation near the oceans contains more SO_4^{2-}, Cl^-, Na^+, and Mg^{2+}

than precipitation falling inland because of the presence of "sea spray" in the atmosphere. Wind-blown soil and dust particles can affect precipitation composition locally.

Table 1-3 shows the composition of precipitation at several different locations. Note that the constituent concentrations of rain and snow are very low, indicating the

Table 1-3   Dissolved Constituents of Rain, Snow, and River Water, in mg/L

Constituent	Snow in the Sierra Nevada[1]	Rain in Eastern NC and VA[2]	Snow in Western Massachusetts[3]	Typical River Water[4]
Ca^{2+}	0.0	0.65	0.2	15
Mg^{2+}	0.2	0.14	0.1	4
Na^+	0.6	0.56	0.8	4
K^+	0.6	0.11	0.2	2
HCO_3^-	3.0	2.12	2.0	58
SO_4^{2-}	1.6	0.92	2.9	11
Cl^-	0.2	2.18	3.8	8
NO_3^-	0.1	0.62	4.4	3
Total dissolved solids	4.8	26	40	150

1. Snow, Spooner Summit, U.S. Highway 50, Nevada (east of Lake Tahoe), altitude 7100 ft., Nov. 20, 1958. J.H. Feth, S.M. Rogers, and C.E. Robertson, Chemical Composition of Snow in the Northern Sierra Nevada and Other Areas, U.S. Geological Water Supply Paper 1535J, 1964.
2. Average composition of rain from August 1962 to July 1963 at 27 points in North Carolina and Virginia. A.W. Gambell and D.W. Fisher, Chemical Composition of Rainfall, Eastern N. Carolina and Southeastern Virginia: U.S. Geological Water Supply Paper 1535K, 1964.
3. Snow, Berlin Mountain, U.S. Highway 2, Massachusetts, altitude 2,300 ft., Feb. 1990. Data from Environmental Science class of Williams College.
4. R. O. Gill, *Chemical Fundamentals of Geology, Unwin Hyman Ltd.* (London, 1989), p. 92.

remarkable efficiency of the evaporation process that produces "fresh water" from seawater. Chemical reactions between rain and snow and the minerals on the earth's surface further influence the concentration of dissolved species in water when the acidic water dissolves substances in the rocks and soil. Because the land surface composition, the extent and nature of plant coverage, and the concentration of atmospheric gases can vary from one location to another, the concentration of dissolved substances in water can be significantly different in different locales.

Effects of Human Activities on Water Quality

Human activities also contribute to the concentrations of dissolved substances in natural waters, particularly in heavily populated areas. These activities also affect the quality of the water delivered to domestic and industrial end-users. The major human activities that affect water quality are human waste disposal through treated or untreated wastewater, agricultural activities, mining, power generation, and industrial processing. The high concentration of people in urban environments ensures that even small sources of pollution from one individual's activities add up to substantial amounts of pollution for an entire city of people. Substances that arise from human activities are called **anthropogenic** pollutants (Figure 1-2).

Figure 1-2:   Human impacts on water quality are significant. Many activities, from farming to manufacturing to mining, contribute pollutants to our waterways.

Human wastewater disposal into rivers and streams is the single largest source of pollutant discharge to waterways. Untreated human waste (uncommon now in the U.S.) contains bacteria, viruses, organic matter, and high concentrations of chloride, nitrates, and phosphates. Organic matter (sugars, proteins, fats, plant lignins, etc.) contained in wastewater is referred to as **biochemical oxygen demand (BOD)**. When waste containing high BOD is discharged into a stream, bacteria in the receiving water will metabolize the organic matter and use up oxygen in the process. For aquatic organisms, this can be disastrous because all of the available oxygen can be consumed by this process. Wastewater from sewage treatment plants also contains high levels of **total suspended solids (TSS)**, small particles that do not settle out in the treatment process, as well as high levels of **total dissolved solids (TDS)**, which mostly consist of ionic substances such as salts, acids and bases, and nitrogen- and phosphorous-containing ions such as nitrate, ammonia, and phosphate. Treated wastewater will contain many fewer bacteria and viruses, and most of the organic matter, nitrate, and phosphate will have been removed. However, the concentration of chloride in treated water is higher because wastewater treatment does not remove much chloride and the chlorination process used to disinfect wastewater actually *adds* chloride to the water.

Agricultural activities involve the use of fertilizers composed of inorganic nitrate, phosphate, and sulfate salts. Because these salts are water soluble, a heavy rainstorm after an application of fertilizer may contaminate waterways with runoff from the fields. Pesticides applied to crops also wash off with rainwater and contaminate streams, rivers, and even groundwater. Animals such as cows, pigs, and chickens are usually concentrated in small areas, and their wastes (almost always untreated) contain high levels of BOD and often run directly into rivers and streams. Animal waste contains high concentrations of organic matter, bacteria, viruses, chloride, nitrates, and phosphates. In areas with shallow groundwater, agricultural polluntants from the surface can **leach**, or pass through the soil and contaminate the aquifer below.

Mining of coal or metal ores creates large quantities of suspended solids (TSS) and often results in the exposure of sulfide-containing minerals, particularly pyrite (FeS_2), to air. Oxidation of sulfide (S^{2-}) to sulfate (SO_4^{2-}) occurs readily and is catalyzed by bacteria. The net result of this reaction is the production of sulfuric acid.

$$2\,FeS_2\,(s)\ +\ 2\,H_2O\ +\ 7\,O_2\ \longrightarrow\ 2\,Fe^{2+}\ +\ 4\,SO_4^{2-}\ +\ 4\,H^+$$

Urban runoff is also a significant, but difficult to eradicate, source of pollution in highly populated areas. During rainstorms, oil, grease, animal feces, and heavy metals from automobile wear and tear wash off the streets and into the nearest stream. In northern climates, road salt (as either $NaCl$ or $CaCl_2$) used for melting snow is a significant source of sodium (or calcium) and chloride in surface waters.

Table 1-4   Major Anthropogenic Pollutant Sources

Industry	TSS (million lbs. per year)	TDS (million lbs. per year)	BOD (million lbs. per year)	Nitrogen (million lbs. per year)	Phosphorus (million lbs. per year)	Heavy Metals (million lbs. per year)	Total (million lbs. per year)
Municipal waste-water treatment plants	3,850	30,255	3,800	814	73.9	9.3	38,800
Organic chemicals	144	36,540	108	41	1.4	3.6	36,840
Power plants	1,166	18,418	-----	-----	-----	24.4	19,610
Pulp and paper mills	782	16,826	530	< 7	< 1.2	< 0.1	18,150
Foods and beverages	91.9	7,420	54.8	12.3	4.7	-----	7,584
Petroleum refining	< 50	2,390	< 25	15.5	1.5	6.0	2,490
Feedlots	422	< 1300	96	40	21.8	-----	1,880
Iron and steel mills	254	1,324	37.8	-----	-----	7.6	1,623

Source: Adapted from F. Van der Leeden, F. L. Troise, and D. K. Todd, *The Water Encyclopedia* (Lewis Publishers, Boca Raton, FL, 1991), p. 503. Data taken from 1981 Council on Environmental Quality Environmental Trends Report.

Manufacturing and processing industries contribute significant amounts of contaminated water to waterways. Table 1-4 lists some common pollutants contributed by these industries. Industry discharges are regulated and permitted by the National Pollution Discharge Elimination System (NPDES) permitting system; however, this does not prevent all pollutants from being dumped into a river or stream.

Table 1-5   Common Water Pollutants from Industrial Sources

Industry	Pollutants
Food and beverage processing	BOD, TSS, nitrate, chloride, phosphate
Petroleum refining	heavy metals, organic sulfur compounds, salts
Paper manufacturing	acid, base, chloride (from bleaching), BOD
Pharmaceuticals	TSS, salts, nitrogen
Chemical manufacturing	TSS, TDS, BOD, nitrogen, phosphorus, heavy metals, organic compounds, acid, base
Steel-making	heavy metals, acid, base, lime ($Ca(OH)_2$), TSS, BOD
Textiles	TSS, colored organic compounds, base, heat
Leather goods	TSS, chromium, TDS, sulfides, BOD
Power plants	heavy metals, TSS, TDS, heat
Mining	sulfates, acid, heavy metals, TSS, TDS
Electroplating	heavy metals, acid

Source: Adapted from F. Van der Leeden, F. L. Troise, and D. K. Todd, *The Water Encyclopedia* (Lewis Publishers, Boca Raton, FL, 1991), pp. 503 and 547-548.

Developing Ideas

Table 1-3 on page 8 gives the composition of rain and snow in different areas of the U.S. and an average composition of river water.

1. Study the table carefully with a classmate and note any obvious trends in the data.

2. Speculate on the reasons for the differences in the concentrations of the dissolved ionic species in rain water vs. river water. Share your ideas with the class as a whole.

3. Someone has sent you two water samples to analyze, one from a lake in the Adirondack Mountains in upstate New York, and the other from Lake Powell in Utah. Unfortunately, the labels have fallen off of the sample bottles. You re-label the bottles **A** and **B** and proceed with the analysis to find the profile of constituents given in Table 1-6. What additional information will you need to know to determine which sample is which?

Table 1-6   Chemical Profile of Samples A and B

Constituent	Sample **A**	Sample **B**
HCO_3^-	35 mg/L	263 mg/L
SO_4^{2-}	4 mg/L	52 mg/L
Cl^-	8 mg/L	22 mg/L
NO_3^-	7 mg/L	2 mg/L
Na^+	12 mg/L	35 mg/L
K^+	6 mg/L	10 mg/L
Ca^{2+}	22 mg/L	150 mg/L
Mg^{2+}	8 mg/L	25 mg/L
Fe^{2+}/Fe^{3+}	0.3 mg/L	0.05 mg/L

Working with the Ideas

The following problems will help you understand the issues in more depth.

4. Consider a body of water (lake, river, stream, ocean, underground aquifer) near *your* city or town. What human activities contribute to raising the levels of dissolved substances in this body of water? What substances are added to the water by these activities?

5. Pick two of the human activities you listed in question 1 above and design a plan for reducing the contribution of dissolved substances from these activities.

6. There is a belt of limestone rock (composed of calcite and aragonite minerals—see Table 1-2 on page 7) that runs through the Appalachian Mountains in Virginia. The area is riddled with caves containing spectacular formations of stalactites and stalagmites. What major ionic components would you expect to see in the water from this area, and what chemical compound(s) do you think the stalactites and stalagmites are mainly composed of?

7. The rocks in the Sierra Nevada mountains are comprised mostly of granite (composed of feldspars, quartz, and olivine). What dissolved substances would you expect to find in the runoff water from the Sierra Nevada?

8. A farmer applies a fertilizer, ammonium nitrate, NH_4NO_3, to his fields. Ammonium nitrate is a water-soluble ionic substance. Read over the information on the hydrologic cycle and suggest several pathways by which the ammonium nitrate could reach the ocean. Speculate on why only a fraction of the applied ammonium nitrate actually ends up in the ocean.

9. Table 1-7 gives a profile of the chemical composition of a ground water. Compare it to the profile given for average river water (Table 1-3) and explain the differences.

Table 1-7 Concentration of Constituents in a Groundwater

Ion	Concentration (mg/L)
Ca^{2+}	95
Mg^{2+}	34
Na^+	9
K^+	1
$Fe^{2+/3+}$	0.8
HCO_3^-	350
SO_4^{2-}	85
Cl^-	10
NO_3^-	15

10. A group of concerned residents suspects a local factory of illegally dumping pollutants into the river that runs through their town, but the factory says the pollutants are naturally occurring and not from their operation. Design a plan to help the residents determine whether the factory is responsible for the pollution or not.

Exploration 1D

How can we obtain a quantitative profile of the ionic constituents in a water supply?

Creating the Context

Why is this an important question?

In Exploration 1C, we saw that the dissolved constituents in natural waters come from a variety of sources. The composition of the rocks and minerals in an area governs the concentrations of ionic substituents, and the dissolved gases in precipitation affect how much of the minerals dissolve when it rains or snows. Ionic substances transported in the atmosphere and human activities contribute additional dissolved substances to a water source. While there are many different types of pollutants, this Exploration will focus on *ionic* compounds, all of which produce charged species, **cations** (positively charged ions) and **anions** (negatively charged ions) when dissolved in water.

When designing a treatment plan for a water supply, it is useful to have a quantitative measure of the relative concentrations of cationic and anionic constituents present. If the balance between cations and anions is different than what would normally be predicted based on the geology of the area, the water supply may have additional contaminants that were missed in the analysis. This information can point out pollution problems very quickly and allow water treatment engineers to find the source of the pollutant and take steps to remove it. In this Exploration, we will learn the methods used to compare the ratio of anions to cations and answer the question, *How can we obtain a quantitative profile of the ionic constituents in a water supply?*

Preparing for Inquiry

BACKGROUND READING

Ionic Substances Found in Natural Waters

The result of the interaction of precipitation (sometimes acidic) with rocks, minerals, sea spray, and dust is the addition of a variety of dissolved substances to water supplies in lakes, rivers, streams, and reservoirs. The *major* natural constituents of fresh terrestrial surface water (typical concentrations in the range 1.0-1,000 mg/L) are sodium (Na^+), calcium (Ca^{2+}), magnesium (Mg^{2+}), potassium (K^+), bicarbonate (HCO_3^-), sulfate (SO_4^{2-}), chloride (Cl^-), and silica (SiO_2). Secondary constituents of fresh water, with typical concentrations in the range of 0.01-10.0 mg/L, include iron (Fe^{2+}), manganese (Mn^{2+}), carbonate (CO_3^{2-}), nitrate (NO_3^-), fluoride (F^-), and borate (BO_3^{3-}, as boric acid, H_3BO_3). Appendix 1A describes the major sources of ions, their average concentrations in natural fresh waters, and their effects on water quality and water use.

In addition to the species described above, fresh water may contain a number of trace ionic species such as Cu^{2+}, Zn^{2+}, $H_2PO_4^-/HPO_4^{2-}$, Pb^{2+}, Sr^{2+}, and Ba^{2+}. Surface waters may also contain organic compounds from decomposing plants and animals, suspended solids such as clays, and microorganisms. All of these components must be considered when treating water for municipal use.

Types of Water Supplies

Public water supplies in the United States can be classified into three general categories based on the different rock types in the basins from which they originate. The first general type (Type A, Table 1-8) is surface water originating in basins where the major rocks are granite, composed of the minerals quartz, the feldspars, hornblende, biotite, forsterite, and olivine (see Table 1-2 on page 7). Type A waters contain very small amounts of dissolved minerals (usually not more than 30 mg/L of TDS) because these minerals are not very water soluble. The water supplies of New York City (from the Catskill Mountains), San Francisco and Oakland (from the Sierra Nevada Mountains), and Seattle (from the Cascade Mountains) are of this type (Type A, Table 1-8). These waters are "soft" because they contain very low concentrations of Ca^{2+} and Mg^{2+}.

The second general type of surface water (Type B, Table 1-8) originates from basins containing non-granitic rock. This type of water includes the Great Lakes (except Lake Superior) and is widely used as public water supplies for cities such as Chicago, Cleveland, Buffalo, Niagara Falls, Detroit, and Milwaukee. This water is of intermediate "hardness" and total mineral content.

The third general type of water is groundwater originating in basins where the sediments are predominantly limestone, with calcite, aragonite, and dolomite as the major minerals composing the rock. These waters occur widely in the Midwest United States and are typified by the water supply of Dayton, Ohio (Type C, Table 1-8). These waters generally have higher dissolved mineral concentrations than surface waters because of the long and intimate contact between the water and the rocks and soils of the aquifer. In addition, soil microorganisms may produce carbon dioxide that further increases mineral dissolution. These waters are called "hard" because of the high Ca^{2+} and Mg^{2+} concentrations and often require softening (removal of Ca^{2+} and Mg^{2+}) to make them acceptable for domestic and industrial uses.

Table 1-8   Typical Analyses of Surface and Ground Waters in the United States

Constituent	Type A[1] (mg/L)	Type B[2] (mg/L)	Type C[3] (mg/L)
Ca^{2+}	13	36	92
Mg^{2+}	2.7	8.1	34
Na^+	4.5	6.5	8.2
K^+	1.4	1.2	1.4
$Fe^{2+/3+}$	0.015	0.02	0.09
HCO_3^-	27	119	339
SO_4^{2-}	16.2	22	84
Cl^-	10.3	13	9.6
NO_3^-	0.7	0.1	13
$SiO_2(aq)$	2.4	12	10
Total Dissolved Solids	68	165	434
Total Hardness as $CaCO_3$	44	123	369

1. East Bay Municipal Utility District, Oakland, California. Average data for 1994.
2. Niagara River, Niagara Falls, N.Y.
3. Well water, Dayton, Ohio.

Electroneutrality of Aqueous Solutions

An overall measure of the concentration of dissolved species in water is the **total dissolved solids (TDS)**. This parameter is usually reported in milligrams of solids per liter of water. In rainwater, the TDS is very low, approximately 10–50 mg/L. Surface waters that are freshwater usually have a TDS content of 100–1,500 mg/L, and seawater has a TDS content of 30,000–40,000 mg/L.

While an analysis of TDS gives a measure of the *total* amount of dissolved species in solution, it is more informative to have a measure of specific ion concentrations. It is then possible to compare the ratio of moles of cations to moles of anions and to look for any discrepancies from the expected ratio. It is important to remember that all solutions are electrically neutral, which means that the total number of positive charges in a solution must be equal to the total number of negative charges in the solution. Ions of one charge cannot be added to, formed in, or removed from a solution without the addition, formation, or removal, of an equal number of ions of the opposite charge. This is the principle of **electroneutrality**. The charge balance, or electroneutrality, equation states that, for a given volume of solution:

Total moles of positive charge = total moles of negative charge

In practical terms, one can obtain the number of moles of positive charge per liter of each dissolved cation by multiplying the charge on each cation times the concentration of the cation in solution in moles per liter. If one sums the moles of positive charge per liter for all cations, the result is the total moles of positive charge per liter of solution. For example, the calcium ion, Ca^{2+}, carries a charge of 2+. Therefore, a solution containing 1 mole of Ca^{2+} per liter contains:

$$\frac{1 \text{ mol of ions}}{L} \times \frac{2 \text{ mol of charge}}{\text{mol of ions}} = \frac{2 \text{ mol of charge}}{L}$$

To simplify the units used in this calculation, the moles of charge per liter is often referred to as **equivalents per liter** or **eq/L**. In this context, an **equivalent** is simply the moles of charge for that particular species. For example, a solution containing 0.1 moles of Ca^{2+} per liter has 0.2 *equivalents* of Ca^{2+} per liter, and a solution containing 0.1 moles of K^+ per liter has 0.1 *equivalents* of K^+ per liter. For the purpose of determining electroneutrality, it is necessary to know solute concentrations in units of molarity or moles per liter.

Table 1-9   Anion-Cation Balance for a Typical Groundwater

	Concentration of ion (mg/L)	Molecular weight of ion (mg/mmol)	Concentration of ion (mmol/L)	Charge per ion	Millimoles of charge per L (meq/L)
Cations					
Ca^{2+}	92	40.1	2.29	2+	4.59
Mg^{2+}	34	24.3	1.40	2+	2.80
Na^+	8.2	23.0	0.36	+1	0.36
K^+	1.4	39.1	0.036	+1	0.036
				Sum	**7.80**
Anions					
HCO_3^-	339	61.0	5.56	-1	5.56
SO_4^{2-}	84	96.1	0.87	-2	1.75
Cl^-	9.6	35.5	0.27	-1	0.27
NO_3^-	13	62.0	0.21	-1	0.21
				Sum	**7.79**

An Example of Anion-Cation Balance

The charge balance equation has significant practical application in water analysis. Since all waters must be electrically neutral, an accurate analysis of the charged species in a water should produce a result in which the total moles of positive charge is equal to the total moles of negative charge. In an acceptable water analysis, this condition must be met to within ±3 percent. Larger deviations indicate either analytical errors or an overlooked charged species. An example anion-cation balance is given in Table 1-9 for a typical groundwater. An error analysis for the charge balance shown in Table 1-9 is calculated by dividing the absolute value of the difference between the total number of positive charges and the total number of negative charges by the average of these two values and converting to percent, as shown below:

$$\frac{|7.80 - 7.79|}{(7.80 + 7.79)/2} \times 100\% = 0.12\%$$

This error analysis shows an agreement between positive and negative charges to within 0.12%, indicating that the electroneutrality analysis is acceptable.

The Bar Diagram

The **bar diagram** is used by water treatment engineers to give a graphical representation of the electroneutrality of a solution. The concentration of each species in millimoles of charge per liter (**milliequivalents per liter** or **meq/L**) is represented by a horizontal bar whose total length is proportional to concentration. There are separate, but adjacent, bars for cations and anions. The order of plotting the major cations is Ca^{2+}, Mg^{2+}, Na^+, K^+. The order of plotting the major anions is HCO_3^- (or CO_3^{2-}), SO_4^{2-}, Cl^-. This order of presentation has been adopted because it allows a crude assessment of the types of salts that will precipitate as the water evaporates. An example bar diagram for the groundwater data in Table 1-9 (see Type C, Table 1-8) is given in Figure 1-3. In this Exploration, you will draw bar diagrams and use the principle of electroneutrality to determine whether a water quality analysis is providing sufficient information about the water supply in question.

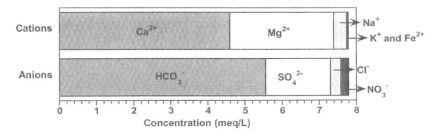

Figure 1-3:   Bar diagram for type C groundwater, from a well in Dayton, Ohio.

Developing Ideas

1. Create a bar diagram for Type B water using the following data. You will first need to calculate the missing values in the table.

	Concentration of ion (mg/L)	Molecular weight of ion (mg/mmol)	Concentration of ion (mmol/L)	Charge per ion	Millimoles of charge per L (meq/L)
Cations					
Ca^{2+}	36	40.1	0.90	2+	1.80
Mg^{2+}	8.1	24.3		2+	
Na^+	6.5	23.0		1+	
K^+	1.2	39.1	0.031	1+	0.031
Fe^{2+}	0.02	55.8		2+	
				Sum	
Anions					
HCO_3^-	119	61.0		1-	
SO_4^{2-}	22	96.1	0.23	-2	0.46
Cl^-	13	35.5		1-	
NO_3^-	0.1	62.0		1-	
				Sum	

2. If a water analysis is to be acceptable, the moles of positive charge must equal the moles of negative charge within ±3%. Is the above analysis for a Type B water acceptable based on this standard?

3. How would the omission of iron (Fe^{2+}) from the water analysis affect the accuracy of the electroneutrality calculation?

4. If you were a water treatment engineer wanting to save time and money on analyses, under what conditions would omitting the iron analysis be a wise thing to do?

5. How would the bar diagram you plotted above look different if the source of the sulfate in the water were sulfuric acid, H_2SO_4, instead of sodium, calcium, or magnesium sulfate (Na_2SO_4, $CaSO_4$, $MgSO_4$)? Sketch this bar diagram.

Working with the Ideas

The following problems will give you more practice with the calculations and interpretation of water quality data.

6. Create a bar diagram for Type A water using the data given in Table 1-9 and compare it to that for Type C water in Figure 1-3. Discuss the major similarities and differences in the two waters. Speculate on the cause of these differences.

7. Create bar diagrams for Type A and Type C water (Table 1-9), omitting the bicarbonate ion (HCO_3^-) from the diagrams. Do an error analysis on the ratio of moles of negative charge to moles of positive charge with this omission. Are the analyses for these two waters acceptable with this omission?

8. Table 1-10 presents water quality data for a stream that runs through a farming community before and after a rainstorm. Calculate the milliequivalents/L of cations and anions in both samples and draw bar diagrams for both data sets. Do these data make sense in terms of the electroneutrality principle? Explain the

data, speculating on the source of any problems. *Hint: Reread the question closely and see Appendix 1A.*

Table 1-10 Analyses of Stream Water Before and After a Rainstorm

Constituent	Before storm	After storm
Ca^{2+}	16.4 mg/L	7.3 mg/L
Mg^{2+}	4.49 mg/L	2.2 mg/L
Na^+	23.8 mg/L	20.1 mg/L
HCO_3^-	62.1 mg/L	29.6 mg/L
SO_4^{2-}	48.2 mg/L	22.6 mg/L
Cl^-	8.1 mg/L	6.2 mg/L

Making the Link

Looking Back: What have you learned?

Chemical Principles

What chemistry experience have you gained?

As you worked through Session 1, you gained experience with some important principles of chemistry. An understanding of these principles is valuable in solving a wide range of real-world problems. Your experience with Session 1 should have developed your skills to do the items on the following lists.

Chemistry of Natural Waters

You should be able to:

- Predict expected dissolved ionic constituents in water supplies based on the local geology and human activities in the area (**Exploration 1C**).

- Recognize the major and minor ionic constituents of natural waters. Major constituents are: sodium (Na^+), calcium (Ca^{2+}), magnesium (Mg^{2+}), potassium (K^+), bicarbonate (HCO_3^-), sulfate (SO_4^{2-}), chloride (Cl^-), and silica (SiO_2). Minor constituents include: iron (Fe^{2+}), manganese (Mn^{2+}), carbonate (CO_3^{2-}), nitrate (NO_3^-), fluoride (F^-), and borate (BO_3^{3-}, as boric acid, H_3BO_3) (**Explorations 1C and 1D**).

Solution Chemistry

You should be able to:

- Calculate concentrations of substances in mg/L or moles/L (**Exploration 1D**).
- Convert concentrations from mg/L to mol/L and the reverse (**Exploration 1D**).
- Convert concentrations from mol/L to moles of charge/L (equivalents/L) (**Exploration 1D**).

- Create a bar diagram that quantitatively displays the concentrations of dissolved ions in a water sample and know how to use these diagrams to determine the accuracy of a water quality analysis (**Exploration 1D**).

You may wish to look up more information on these topics in your introductory chemistry textbook for additional explanation and to see other examples. They will most likely be discussed in a chapter on **Solution Chemistry**.

Thinking Skills

What general skills are you building for your resume?

You have also been developing some general problem-solving and scientific thinking skills that are valued in a wide range of professions.

Decision-making Skills

You should be able to:

- Develop a list of scientific criteria needed to make an informed choice about the best plan to do a job (**Exploration 1B**).

Data Analysis Skills

You should be able to:

- Evaluate and compare sets of similar data to look for trends and differences (**Exploration 1C**).
- Create bar plots to compare data from various sources (**Exploration 1D**).

Problem-solving Skills

You should be able to:

- Develop a step-by-step plan to accomplish a goal (**Exploration 1B**).
- Use a set of regulations to help design specifications for a plan (**Exploration 1B**).

Checking Your Progress

What progress have you made toward answering the Module Question?

Appendix 1A

Ionic Substances in Natural Waters

Table A-1: Principal Chemical Constituents in Water: Their Sources, Concentrations, and Effects Upon Usability

Constituent	Major Sources	Concentration in Natural Water	Effect Upon Usability of Water
Carbonate (CO_3^{2-}) Bicarbonate (HCO_3^-)	Limestone, dolomite	Commonly 0 mg/L in surface water; commonly less than 10 mg/L in ground water. Water high in sodium may contain as much as 50 mg/L of carbonate. Commonly less than 500 mg/L; may exceed 1000 mg/L in water highly charged with carbon dioxide.	Upon heating, bicarbonate is changed into steam, carbon dioxide, and carbonate. The carbonate combines with alkaline earths—principally calcium and magnesium—to form a crustlike scale of calcium carbonate that retards flow of heat through pipe walls and restricts flow of fluids in pipes. Water containing large amounts of bicarbonate and alkalinity is undesirable in many industries.
Sulfate (SO_4^{2-})	Oxidation of sulfide ores; gypsum; anhydrite; industrial wastes	Commonly less than 1000 mg/L except in streams and wells influenced by acid mine drainage—as much as 200,000 mg/L in some brines	Sulfate combines with calcium to form an adherent, heat-retarding scale. More than 250 mg/L is objectionable in water in some industries. Water containing about 500 mg/L of sulfate tastes bitter; water containing about 1000 mg/L may be cathartic.
Chloride (Cl^-)	Chief source is sedimentary rock (evaporites); minor sources are igneous rocks. Ocean tides force salty water upstream in tidal estuaries	Commonly less than 10 mg/L in humid regions; tidal streams contain increasing amounts of chloride (as much as 19,000 mg/L) as the bay or ocean is approached. About 19,300 mg/L in seawater; and as much as 200,000 mg/L in brines.	Chloride in excess of 100 mg/L imparts a salty taste. Concentration greatly in excess of 100 mg/L may cause physiological damage. Food processing industries usually require less than 250 mg/L. Some industries—textile processing, paper manufacturing, and synthetic rubber manufacturing—desire less than 100 mg/L.
Fluoride (F^-)	Amphiboles (hornblende), apatite, fluorite, mica	Concentrations generally do not exceed 10 mg/L in ground water or 1.0 mg/L in surface water. Concentrations may be as much as 1600 mg/L in brines.	Fluoride concentration between 0.6 and 1.7 mg/L in drinking water has a beneficial effect on the structure and resistance to decay of children's teeth. Fluoride in excess of 1.5 mg/L in some areas causes "mottled enamel" in children's teeth. Fluoride in excess of 6.0 mg/L causes pronounced mottling and disfiguration of teeth.
Nitrate (NO_3^-), reported as N	Atmosphere; legumes, plant debris, animal excrement, nitrogenous fertilizer in soil and sewage	In surface water not subjected to pollution, concentration of nitrate may be as much as 5.0 mg/L but is commonly less than 1.0 mg/L. In groundwater the concentration of nitrate may be as much as 1000 mg/L.	Water containing large amounts of nitrate (more than 100 mg/L) is bitter tasting and may cause physiological distress. Water from shallow wells containing more than 45 mg/L has been reported to cause methemoglobinemia in infants. Small amounts of nitrate help reduce cracking of high-pressure boiler steel.
Dissolved solids	The mineral constituents dissolved in water constitute the dissolved solids.	Surface water commonly contains less than 3000 mg/L; streams draining salt beds in arid regions may contain in excess of 15,000 mg/L. Ground water commonly contains less than 5000; some brines contain as much as 300,000 mg/L.	More than 500 mg/L is undesirable for drinking and many industrial uses. Less than 300 mg/L is desirable for dyeing of textiles and the manufacture of plastics, pulp paper, rayon. Dissolved solids cause foaming in steam boilers; the maximum permissible content decreases with increases in operating pressure.
Silica (SiO_2)	Feldspars, ferromagnesium and clay minerals, amorphous silica, chert, opal	Ranges generally from 1.0 to 30 mg/L, although as much as 100 mg/L is fairly common; as much as 4000 mg/L is found in brines.	In the presence of calcium and magnesium, silica forms a scale in boilers and on steam turbines that retards heat; the scale is difficult to remove. Silica may be added to soft water to inhibit corrosion of iron pipes.

Constituent	Major Sources	Concentration in Natural Water	Effect Upon Usability of Water
Iron (Fe)	Natural sources: Igneous rocks such as amphiboles, ferromagnesian, micas, ferrous sulfide (FeS), ferric sulfide or iron pyrite (FeS_2), magnetite (Fe_3O_4); sandstone rocks such as oxides, carbonates, and sulfides or iron clay minerals. Manmade sources: Well casing, piping, pump parts, storage tanks, and other objects of cast iron and steel that may be in contact with the water	Generally less than 0.50 mg/L in fully aerated water. Ground water having a pH of less than 8.0 may contain 10 mg/L; rarely as much as 50 mg/L may occur. Acid water from thermal springs, mine wastes, and industrial wastes may contain more than 6000 mg/L.	More than 0.1 mg/L precipitates after exposure to air; causes turbidity, stains plumbing fixtures, laundry and cooking utensils, and imparts objectionable tastes and colors to foods and drinks. More than 0.2 mg/L is objectionable for most industrial uses.
Manganese (Mn^{2+})	Manganese in natural water probably comes most often from soils and sediments. Metamorphic and sedimentary rocks and mica biotite and amphibole hornblende minerals contain large amounts of manganese.	Generally 0.20 mg/L or less. Groundwater and acid mine water may contain more than 10 mg/L. Reservoir water that has "turned over" may contain more than 150 mg/L.	More than 0.2 mg/L precipitates on oxidation; causes undesirable tastes, deposits on foods during cooking, stains plumbing fixtures and laundry, and fosters growths in reservoirs, filters, and distribution systems. Most industrial users object to water containing more than 0.2 mg/L.
Calcium (Ca^{2+})	Amphiboles, feldspars, gypsum, pyroxenes, aragonite, calcite, dolomite, clay minerals	As much as 600 mg/L in some western streams; brines may contain as much as 75,000 mg/L.	Both calcium and magnesium combine with bicarbonate, carbonate, sulfate and silica to form heat-retarding, pipe-clogging sale in boilers and in other heat-exchange equipment. Calcium and magnesium combine with ions of fatty acid in soaps to form soap suds; the more calcium and magnesium, the more soap required to form suds. A high concentration of magnesium has a laxative effect, especially on new users of the supply.
Magnesium (Mg^{2+})	Amphiboles, olivine, pyroxenes, dolomite, magnesite, clay minerals	As much as several hundred mg/L in some western streams; ocean water contains more than 1000 mg/L and brines may contain as much as 57,000 mg/L.	
Sodium (Na^+)	Feldspars (albite); clay minerals; evaporites, such as halite (NaCl) and mirabilite ($Na_2SO_4 \cdot H_2O$); industrial wastes	As much as 1000 mg/L in some western streams; about 10,000 mg/L in sea water; about 25,000 mg/L in brines.	More than 50 mg/L sodium and potassium in the presence of suspended matter causes foaming, which accelerates scale formation and corrosion in boilers. Sodium and potassium carbonate in recirulating cooling water can cause deterioration of wood in cooling towers. More than 65 mg/L of sodium can cause problems in ice manufacture.
Potassium (K^+)	Feldspars (orthoclase and microcline), feldspathoids, some micas, clay minerals	Generally less than about 10 mg/L; as much as 100 mg/L in hot springs; as much as 25,000 mg/L in brines.	

Reproduced with permission from F. Van der Leeden, F. L. Troise, and D. K. Todd, *The Water Encyclopedia* (Lewis Publishers, Boca Raton, FL, 1991), pp. 422–423.

Analysis

How do you determine how much of a constituent is in a water supply?

Exploration 2A: The Storyline

Creating the Context

Why is this an important question?

In the laboratory parts of this module, your mission will be to develop a purification procedure for a water sample that is out of range of the drinking water standards. Beginning in this Session and continuing in later laboratory explorations, you will work in teams to bring this water sample into compliance. We will focus on the three most common contaminants that water treatment plants must remove before distributing the water to users: fluoride, water hardness ions (calcium and magnesium), and iron.

You will be given an impure water sample with a single contaminant above the Maximum Contaminant Level (MCL), one of either: 1) fluoride, 2) the hardness ions calcium and magnesium, or 3) iron. These samples are representative of water supplies in various parts of the country. Over the next several weeks, you will work with your teammates to design a procedure for reducing the concentration of your specific contaminant to an acceptable level: that is, below the MCL (see http://www.epa.gov/waterscience/drinking/standards/ for the MCLs for these contaminants). Please be aware that there is no single correct method for removing a contaminant. Many approaches may work, and every sample will be a bit different. Your performance will be evaluated on how well you think through the purification procedure and on how well you understand the chemistry behind the purification and analytical methods.

In this Session, each person will begin by learning the analytical technique for measuring the concentration of one of these ions and will determine the concentration of the ion in his/her *own* sample. All samples will be somewhat different. The goal of this Session is to learn the answer to the question, *How do you determine how much of a constituent is in a water supply?*

Developing Ideas

What exactly does it mean to "analyze a sample" for a particular component or property? To explore this concept in a general sense, work with one or two other students to answer *one* of the first three questions. Be sure that each of the three questions is answered by at least one group in the class.

For Questions 1-3, devise an analytical plan to determine the answer and describe the tools you would need for the analysis.

1. A member of the Olympic organizing committee is in charge of drug testing for the Olympic athletes. She needs to determine:

    a. What percent of the group of athletes uses some type of performance-enhancing drug?

    b.   What types of drugs are used by the athletes?

    Specifically describe the tools she would need for the analysis.

2.   You are given a mixture of five different kinds of candies and are asked to find out how many of each kind you have. You are blindfolded.

3.   Soil is composed of particles of many different sizes. Soil scientists often need to determine how many grams of each particle size is present per 100 g of sample.

4.   Discuss your results from Problems 1–3 with the class as a whole. What common themes emerge in the process of developing an analytical plan?

5.   Your instructor will now do a chemical analysis on three different solutions containing equal concentrations of Na^+, Ca^{2+}, and Mg^{2+}, respectively. Observe the reactions and discuss the following questions:

    a.   Will the analysis allow accurate determination of the concentration of Na^+? Ca^{2+}? Mg^{2+}?

    b.   Under what conditions will the analysis be accurate? What factors might cause it to be inaccurate?

6.   Your instructor will show you three solutions containing different concentrations of copper. How can you distinguish them?

Looking Ahead

During the first laboratory period you will form a group with two other students with whom you will work for the duration of the module. Each group should obtain a contaminated water sample from the instructor. Be sure to note the sample number in your laboratory notebook. Your team will follow the instructions in one of either Exploration 2B, 2C, or 2D for determining the concentration of the specific contaminant in your sample, either fluoride, water hardness, or iron. You will need to carry out your analysis at least twice and average the numbers with other groups to obtain a more reliable measure of the concentration of the substance in the sample.

- **Exploration 2B:** How do you determine how much fluoride is in a water supply?

- **Exploration 2C:** How do you determine how much calcium and magnesium are in a water supply?

- **Exploration 2D:** How do you determine how much iron is in a water supply?

Each student will then determine the total alkalinity of his/her sample and the concentration of total dissolved solids (TDS) in the sample. Each student in the group need only do a single analysis for both of these measurements. Average the three values obtained by the group members for total alkalinity and TDS to obtain more reliable measures.

- **Exploration 2E:** How do you determine the total alkalinity in a water supply?

- **Exploration 2F:** How do you determine the concentration of total dissolved solids (TDS) in a water supply?

Exploration 2B

How do you determine how much fluoride is in a water supply?

Creating the Context

Why is this an important question?

Fluoride in the soil can be found in the form of the minerals fluoroapatite $(Ca_{10}(PO_4)_6F_2)$ and fluorite (CaF_2). Although fluorine is the 17th most abundant element in the earth's crust, a majority of the water supplies that are tested contain fluoride at levels well below EPA limits. It is more the solubility of the fluoride-containing minerals than their abundance that determines the amount of fluoride present in the water. Because of the presence of soluble fluoride-containing minerals, some communities have no choice but to defluoridate their water. One such community is Desert Center, California. This city makes use of a 1.5 million gallon per day defluoridation plant to reduce the amount of fluoride in their water from 8 ppm to 1 ppm.

Such remediation is important, considering the adverse health effects of excess fluoride, such as mottling of teeth and impaired skeletal formation. However, total removal of fluoride is also not desirable. Communities that have low levels of fluoride (<1 ppm) actually *add* fluoride to their water. The EPA endorses the controlled fluoridation of waters that are deficient in fluoride, since studies have shown that low levels of fluoride prevent the formation of dental cavities. Normally, waters are fluoridated to levels of 1 to 2 ppm, depending on temperature, since it is assumed that more water is consumed at higher ambient temperatures.

What do you already know?

As you learned in Exploration 2A, the key feature of any reliable analytical technique is that the analysis be *specific* for the analyte (the substance of interest), even in the presence of other substances. By the end of this Exploration, you will have learned the answer to the question *How do you determine how much fluoride is in a water supply?*

Preparing for Inquiry

BACKGROUND READING

The concentration of fluoride and other ions in aqueous solution can be determined using an **ion-selective electrode**, abbreviated **ISE**. Ion-selective electrodes are instruments that respond selectively to a specific ion of interest in a solution. The response is measured as a small voltage across the solution using a voltmeter. The measured voltage, or electrical potential, is a function of the concentration of the specific ion in solution.

Ion-selective electrodes are used in conjunction with a reference electrode. The two electrodes are connected via electrical wire to the voltmeter (see Figure 2-1). When the electrodes are immersed in a sample solution, a circuit is formed which can conduct electric current. The ISE for the measurement of fluoride consists of an epoxy sheath into which a solution of known fluoride concentration and a lanthanum fluoride (LaF_3) crystal are encased. The lanthanum fluoride crystal responds to differences in the concentration of fluoride ion between the sample solution and the inner electrode solution by generating a small electrical current. The electrical current generated in the circuit is detected by the voltmeter and registered as an electrical potential. The magnitude of the electrical potential is proportional to the concentration of the fluoride ion in solution.

Figure 2-1:   The combined electrode contains both the fluoride-selective electrode and the reference electrode in a single housing. The voltmeter provides a measure of electrical potential, which depends on the concentration of fluoride ion in the solution.

The reference electrode provides an unvarying electrical potential regardless of the concentration of ions in the solution into which it is immersed. This electrode provides a baseline electrical potential against which the voltage generated by the ion-selective electrode can be measured.

Before the sample solution can be accurately analyzed, the ISE must be standardized against solutions of *known* fluoride ion concentration. Based on the voltages measured for each standard solution, a linear relationship can be derived relating log [F⁻] to the measured voltage, E_{cell}. The ideal relationship between the concentration of fluoride ions and the measured electrical potential can be characterized by the following equation at 25°C:

$$E_{cell} \text{ (mV)} = \text{constant} - 59.2 \log [F^-]$$

The constant is specific to the type of ISE and reference electrode being used. In reality, the factor of 59.2 (the slope of the line) may be slightly different because of the presence of other ions in solution. In the example shown in Figure 2-2, the slope of the line is 50.4.

The solution to be analyzed is maintained in a pH range between 5.2 and 5.7 to reduce the concentrations of interfering compounds such as hydroxide ions and hydrofluoric acid. A buffer solution that stabilizes pH is typically added to the sample for this purpose. A cation-complexing reagent is also added to the sample to reduce the concentrations of cations such as Fe^{3+}, Al^{3+}, and Ca^{2+} which actively bind to fluoride ions.

The fluoride ISE responds differently to solutions in which the total concentration of ionic substances, the **ionic strength**, is significantly different. For example, if two solutions have identical fluoride concentrations, but one solution is more concentrated in *other* ions, the electrode would give different readings for the fluoride concentration because of the interference of the other ions. To avoid errors in measuring solutions with an ionic strength different from the standards, the ISE method for fluoride analysis calls for conditions of high ionic strength. This minimizes any change in electrode response due to the concentration of other ions in the solution.

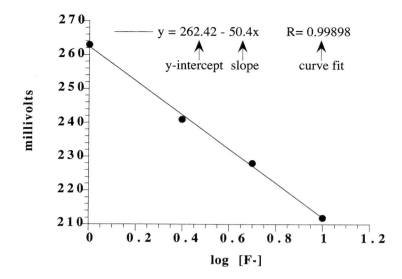

Figure 2-2: A calibration curve for the fluoride-selective electrode is created by measuring the electrical potential of several solutions of known fluoride concentration and plotting log[F⁻] vs. millivolts.

PRELAB QUESTIONS

Read the **Background Reading** section on fluoride on pages 26–27 and the **Procedures** for determining fluoride on pages 28–30. Answer the questions below in your laboratory notebook before arriving in the laboratory.

1. What chemical compounds are the source of fluoride in natural waters? With what minerals are they associated?
2. What problems are associated with too much fluoride in a water supply? With too little fluoride in a water supply?
3. What is the purpose of adding the buffer solution before doing the fluoride analysis?
4. What type of mathematical transformation will you need to carry out before you make your calibration curve?

Developing Ideas

PROCEDURES FOR DETERMINING FLUORIDE

In this experiment, you will determine fluoride concentration by the use of an ion-selective electrode. Background information on the technique can be found on pages 26–27.

The fluoride and reference electrodes are extremely delicate and costly instruments and are especially fragile on the bottom. Do not handle the bottom of the electrodes, and do not bump the electrode against the bottom of the beaker.

Standardizing the Fluoride ISE

Before you can obtain accurate measurements, the electrode/meter combination must be standardized using solutions with a known fluoride concentration. There are four standards; 1.0 ppm, 5.0 ppm, 10.0 ppm, and 15.0 ppm.

note   All glassware used in this experiment should be very clean and well rinsed with distilled water. If the glassware will not come clean with normal procedures, use the procedures in Appendix 2A.

1. Prepare a table in your lab notebook with concentration in one column, log of the concentration in another, and millivolts in a third. Use this table to record your measurements.

2. Turn on the voltmeter by pressing the on/off button and be sure the display reads out in millivolts (mV). Note the room temperature.

3. The standard procedure for transferring the electrode from one solution to another is as follows: Remove the fluoride electrode from the solution and rinse it *well* with deionized water, using a wash bottle. Carefully blot it dry with a Kimwipe™ and place it in the new solution.

4. The standards will already be prepared and will be near the meter in 50 mL beakers. Place the beaker containing the lowest concentration standard on the center of the stir plate, and start the magnetic stir bar so it stirs slowly. **VERY GENTLY** lower the (clean) electrodes into the solution, being careful to keep the electrodes from hitting the stir bar.

5. Wait for the reading to stabilize (which may take a minute or so), and record the millivolt reading on the meter.

6. Repeat this procedure (steps 3–5) for all four standards, working from lowest to highest concentration.

7. *Before you leave the lab*, make a plot of log [F⁻] vs. millivolts. You should obtain a straight line. If not, repeat the calibration procedure. Determine the y-intercept and the slope of the line.

note   The ion-selective electrode needs to be calibrated on a regular basis, at least once every hour. For the most reliable data, calibrate the electrode *every time you use it*.

Determining Fluoride in a Sample

1. Each member of your team should carry out one analysis of the original impure water sample. As you work on removing the contaminant or **remediating** the sample (see Session 6), you will use the same analytical procedure. For the analysis of the remediated sample, a *single* analysis by one team member will be sufficient.

2. To test the sample, you must first add a Total Ionic Strength Adjustment Buffer (TISAB) to the solution. This solution regulates the amount of total ions in the solution to give the fluoride electrode uniform operating conditions. Using a clean graduated cylinder, measure equal volumes of the TISAB solution and the sample water you wish to test into a clean 50 mL beaker. There should be at least 15 mL total volume.

3. Add a stir bar and stir the solution slowly, then **GENTLY** lower the (recently calibrated!) electrodes into the beaker, being careful to keep them above the stir bar. Allow the reading to stabilize and note the voltage reading.

4. For more precise measurements (of both the original sample and the final treated sample), be sure your team analyzes at least three replicate samples. The three measurements should differ by less than 5%. If this is not the case, repeat the sample preparation and analysis if time permits.

5. Share your results with your teammates and average the numbers to obtain a more reliable measure of the fluoride concentration in the sample.

 Dispose of the solutions in the container provided.
disposal

Calculations

Calculate the fluoride concentration of your unknown sample using the slope and y-intercept of your calibration curve in conjunction with the equation describing the linear relationship (see Figure 2-2 on page 28 for an example plot). Don't forget that your calibration curve should plot *log* of the fluoride concentration vs. millivolts, not the concentration itself!

Working with the Ideas

The following problems will help you understand the issues in more depth.

1. What is the maximum allowable contaminant level (MCL) for fluoride? Is your sample within or outside the acceptable range? By how much?

2. In your own words, describe the theory behind the electrochemical method used for measuring fluoride.

3. A sample containing 4.0 mg/L of fluoride was tested with two different fluoride-selective electrodes. One of the electrodes gave a reading of 217 mV and the other gave a reading of 232 mV. Explain how this might happen, even if the analysis is working perfectly.

4. Call the water treatment plant for your community to inquire about the amount of fluoride in your local drinking water.

Exploration 2C

How do you determine how much calcium and magnesium are in a water supply?

Creating the Context

Why is this an important question?

The **total hardness** (**TH**) of water is defined as the sum of the concentrations of all cations in water with more than one charge, that is, the **multivalent** cations such as Ca^{2+}, Mg^{2+}, Fe^{2+}, Sr^{2+}, Ba^{2+}, and Al^{3+}. As you discovered in Session 1, because of their natural geological abundance, calcium and magnesium are the major contributors to hardness in natural waters. Common minerals containing Ca^{2+} and Mg^{2+} include gypsum ($CaSO_4$), calcite ($CaCO_3$), magnesite ($MgCO_3$), and dolomite ($CaMg(CO_3)_2$).

When too many calcium and magnesium ions are present in water, they may cause undesirable chemical reactions that interfere with water use. For example, if you live in a region that has "hard" water (that is, water with many calcium and magnesium ions), you may have noticed that your soap does not lather well. This occurs because Ca^{2+} and Mg^{2+} present in the water react with soap molecules to form an insoluble complex that precipitates out of solution. We know this insoluble precipitate as "soap scum." The effect of this reaction is to remove soap molecules from solution and reduce the soap's capacity to solubilize dirt!

$$M^{2+} (aq) + Na^{+-}OOC\text{-}(CH_2)_{10}\text{-}CH_3 (aq) \longrightarrow M[OOC\text{-}(CH_2)_{10}\text{-}CH_3]_2 (s)$$

hardness ion   **soap**   **insoluble precipitate**

$M = Ca^{2+}, Mg^{2+}$

Water hardness can cause more significant problems in residential and industrial water heaters, boilers, pipes, and heat exchange systems. When water containing Ca^{2+}, Mg^{2+}, and bicarbonate (HCO_3^-) ions is heated, calcium carbonate ($CaCO_3$) and magnesium carbonate ($MgCO_3$) may precipitate from solution. These precipitates, commonly known as "scale," coat the insides of pipes and heater coils, reducing the efficiency of water flow and heat transfer. They also increase the cost of system operation and maintenance, so it is easy to see why communities want to remove some of the hardness ions from water prior to its use. The process of removing hardness from water is called **water softening**. Waters are classified according to their hardness, as outlined in Table 2-1.

As you learned in Exploration 2A, the key feature of any reliable analytical technique is that the analysis be *specific* for the analyte (the substance of interest), even in the presence of other substances. By the end of this Exploration, you will have learned the answer to the question *How do you determine how much calcium and magnesium are in a water supply?*

Table 2-1   Hardness Classification of Waters

Hardness (mg of $CaCO_3$/L)	Classification
<75	Soft
75-150	Moderately hard
150-300	Hard
>300	Very hard

The convention in water chemistry is to treat water hardness as if it were *all* due to $CaCO_3$, even if the sample contains both Mg^{2+} and Ca^{2+} and even if the counter-ion is not CO_3^{2-}.

Preparing for Inquiry

BACKGROUND READING

The Relationship of Total Hardness to Total Alkalinity

In natural water supplies, calcium and magnesium cations are often accompanied by the ions responsible for **alkalinity**, carbonate (CO_3^{2-}), bicarbonate (HCO_3^-), and hydroxide. Because these species often occur together in natural waters, the ($[Ca^{2+}] + [Mg^{2+}]$) ion concentration in moles per liter is often approximately the same as *half* the alkalinity concentration in moles per liter because of the stoichiometry of the dissolution reaction, shown here:

$$MCO_3\ (s)\ +\ CO_2\ +\ H_2O\ \longrightarrow\ M^{2+}\ +\ 2\ HCO_3^-$$

$$M = Ca^{2+},\ Mg^{2+}$$

	hardness	alkalinity
	1 mole	2 moles

Because of this relationship, the total alkalinity of a water can be a reasonable guide to its hardness, and vice versa. You will learn more about alkalinity in Exploration 2E.

Measuring Water Hardness

Although water hardness includes all multivalent cations, in practice we measure water hardness as the sum of calcium and magnesium ions because they are the most abundant in natural waters. The most common technique for measuring water hardness in the lab involves titrating the sample with a reagent that binds or forms complexes with Ca^{2+} and Mg^{2+}. **Titration** is the dropwise addition of a reagent that reacts essentially completely with the analyte of interest. One of the most effective complexing agents for determining water hardness is **ethylenediaminetetraacetic acid** (abbreviated **EDTA**). The EDTA molecule (shown in Figure 2-3) contains a number of oxygen and nitrogen atoms. These atoms have unshared pairs of electrons that act as bonding sites for cations such as Ca^{2+} and Mg^{2+}. The complex formed by the bonding of EDTA with these ions is called a **chelate**, a name given to any complex that is formed by the binding of a metal ion to more than one atom from the complexing molecule. The effect of chelate formation is to keep the cations in solution and prevent them from participating in other reactions.

In this laboratory analysis, you will be using the disodium salt of EDTA, $Na_2[EDTA-H_2]$. When disodium EDTA is dissolved in water, it forms a di-anion, represented as $[EDTA-H_2]^{2-}$. This di-anion reacts with any metal ion, M^{2+}, with a 2+ charge or greater, according to the reaction shown in Figure 2-3. When the $[EDTA-H_2]^{2-}$ di-anion reacts with a metal cation, a very stable complex forms in which two nitrogen atoms and four oxygen atoms are bound directly to the metal atom (see Figure 2-3). The stoichiometry of the reaction is such that one mole of

EDTA reacts with one mole of a metal cation (Mg^{2+}, Ca^{2+}) to form one mole of the chelate complex.

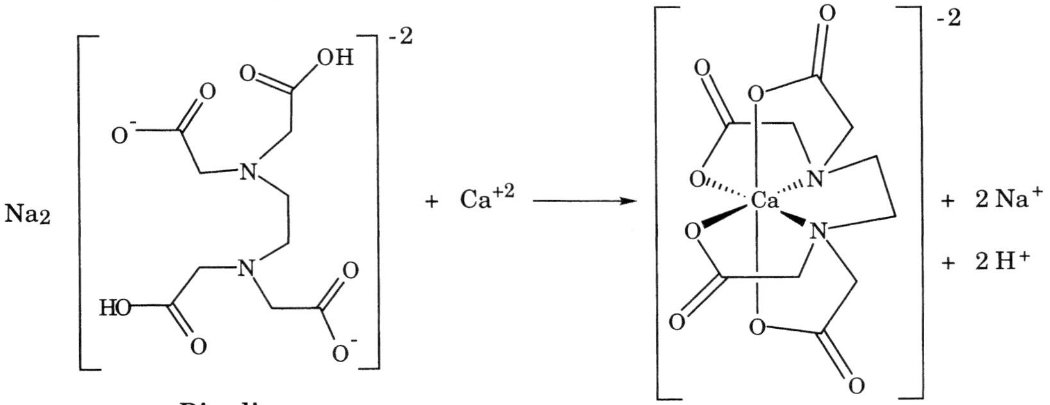

Disodium ethylenediaminetetraacetic acid Na$_2$ (EDTA)

[Ca(EDTA)]$^{-2}$ complex

Figure 2-3: EDTA reacts with metal ions to form a tightly bound chelate complex which is more water soluble than the aquated metal ion.

How do you know when you have added enough EDTA to your solution to complex all of the cations? An indicator dye, called Eriochrome Black T (EBT), is added to the solution prior to titration. This dye forms a weak complex with some of the calcium and magnesium in solution, causing it to turn wine red. When EDTA is added to the solution, it first binds to the free Ca^{2+} and Mg^{2+} in solution. Once the free Ca^{2+} and Mg^{2+} in solution have been complexed, EDTA begins to successfully compete with the EBT for these ions. While EBT bound to calcium and magnesium is wine red in color, unbound EBT is blue. A change in the color of the solution from red to blue thus signals that the titration endpoint has been reached and all Ca^{2+} and Mg^{2+} in the sample are now bound to EDTA.

PRELAB QUESTIONS

Read the **Background Reading** section on water hardness on pages 31–33 and the **Procedures** for determining water hardness on pages 34–35. Answer the following questions in your laboratory notebook before arriving in the laboratory.

1. What chemical compounds are the source of hardness in natural waters? With what minerals are they associated?

2. What problems are associated with too much hardness in a water supply?

3. The binding of [EDTA-H$_2$]$^{2-}$ to the hardness ions to form the M-EDTA complex is an equilibrium reaction, as is the binding of the indicator EBT to the hardness ions.

    a. Write out the chemical equations describing the two equilibria.

    b. Which reaction has the larger equilibrium constant? Explain how you know.

4. Briefly outline the procedure you will follow to determine the concentration of water hardness in your sample.

Developing Ideas

PROCEDURES FOR DETERMINING WATER HARDNESS

In this experiment, you will determine the water hardness by EDTA titration. Background information on the technique can be found on pages 32-33.

---

note

All glassware used in this experiment should be very clean and well rinsed with distilled water. If the glassware will not come clean with normal procedures, use the procedures in Appendix 2A.

---

1. Each member of your team should carry out one analysis of the original impure water sample. As you work on removing the contaminant or **remediating** the sample (see Session 6), you will use the same analytical procedure. For the analysis of the remediated sample, a *single* analysis by one team member will be sufficient.

2. Measure 5.0 mL of the water sample into a 250 mL Erlenmeyer flask. Add:

   ✗ 50 mL of distilled water No

   • 3.0 mL of pH 10 buffer

   • 6 drops of Eriochrome Black T indicator.

   • 1 mL of a 1:1 Mg:EDTA solution to sharpen the endpoint (Note: This may already be contained in the pH 10 buffer solution. Check the labels carefully.)

---

note

The volume of the sample required for an optimum titration might not be 5 mL, depending on the hardness of the water. If the titration requires you to refill your buret, you should use less sample and adjust your calculation accordingly. If only a few mL of EDTA titrant are required to reach the endpoint, redo the titration using more sample. Be sure to note the *actual* sample volume in your laboratory notebook and use that value in your calculations.

---

3. Fill your buret with the EDTA solution. Be sure to record the starting volume of the EDTA in the buret to at least two decimal places, using the lines on the buret to estimate the last decimal place. Also note the *exact* concentration of the EDTA as listed on the side of the bottle.

4. Check that the buret tip contains no air bubbles. If it does, allow a few mL of EDTA solution to flow out. You may need to tap the buret to dislodge the air bubbles.

5. Take the initial volume reading from the buret. When you read the buret, always have your eyes at the same height as the liquid and read the value at the bottom of the meniscus to one decimal place (e.g., 10.1) and estimate the second decimal place.

6. Begin to titrate, swirling the flask and using water from your wash bottle to wash down the sides of the flask periodically. If you miss the endpoint, you can add additional sample using a graduated pipet (be sure to note exactly how much you add) and try approaching the endpoint again. An ideal titration will use between 10 and 20 mL of EDTA. (Why is this range ideal?) The final endpoint should be a blue-violet color. You may wish to keep the flask with a good endpoint color to use for comparison with subsequent titrations.

7. For more precise measurements, analyze at least three replicate samples. The three analyses should differ by less than 5%. If this is not the case, repeat the sample preparation and analysis if time permits.

 Dispose of the waste in the container provided.

disposal

Calculations

1. From the volume and concentration of EDTA solution used to titrate the sample, calculate the concentration of total hardness ($Ca^{2+} + Mg^{2+}$) in the sample in millimoles per liter. This is a four-step process:

    *   Convert mL of EDTA solution used to titrate the sample to moles of EDTA.

    *   Use the mole ratio of reactants in the reaction of Ca^{2+}, Mg^{2+} and EDTA to determine the moles of $[Ca^{2+} + Mg^{2+}]$ that must have been present initially.

    *   Use the volume of the sample to obtain the concentration of $[Ca^{2+} + Mg^{2+}]$ in the sample in moles per liter.

    *   Convert moles per liter to millimoles per liter and report your concentration of hardness in millimoles per liter.

2. Remember that the convention in water chemistry is to treat water hardness as if it were *all* due to $CaCO_3$. You'll need to convert the total moles of $[Ca^{2+} + Mg^{2+}]$ per liter that you just calculated to total hardness in mg of $CaCO_3$ per liter. Compare the value you obtained to Table 2-1. Is your water sample soft, moderately hard, hard, or very hard?

Working with the Ideas

The following problems will help you understand the issues in more depth.

1. If you look in the Drinking Water Standards, you will not find a maximum contaminant level (MCL) for water hardness. Why is this so?

2. In your own words, describe the theory behind the titrimetric method used for measuring water hardness.

3. In preparing samples for an EDTA titration, the addition of a Mg:EDTA complex is called for. If the analysis is measuring the amount of calcium and magnesium in the sample, why doesn't this reagent throw off the accuracy of the analysis?

4. A student is carrying out a water hardness analysis as written in the preceding procedures. On addition of just 5 drops of EDTA solution, the titration reaches the endpoint and the color changes from red to blue. Because it is difficult to accurately measure the volume of such a small quantity of titrant, the procedure must be altered. How would you modify the procedure so that more titrant is required for the titration? Be as specific as possible.

Exploration 2D

How do you determine how much iron is in a water supply?

Creating the Context

Why is this an important question?

Iron is a very abundant element, and measurable concentrations of iron often exist in natural waters, particularly in well water. A variety of common minerals contain iron, including hematite (Fe_2O_3), pyrite (FeS_2), magnetite (Fe_3O_4), and some of the silicate minerals such as olivine $(FeMg)SiO_4$, and fayalite, Fe_2SiO_3. Although a high concentration of iron in drinking water will cause no real health problems, communities often wish to remove it from the water supply for aesthetic reasons, since high concentrations of iron can make water appear murky, smell and taste bad, and stain plumbing fixtures.

Iron in water can occur in two oxidation states, Fe^{2+} or Fe^{3+}. Iron (III) is the form of iron found in ordinary rust (iron oxide, Fe_2O_3). This form of iron is quite stable and will not react with oxygen in the air. In contrast, iron (II) is very reactive towards oxygen. If a solution of Fe^{2+} is left exposed to air for several hours, much of the Fe^{2+} will be oxidized to Fe^{3+} and a rusty-colored, gelatinous precipitate of iron (III) hydroxide, $Fe(OH)_3$, will form.

Both oxidation states of iron form very insoluble hydroxides, with $Fe(OH)_3$ being much less soluble than $Fe(OH)_2$. Because of the extreme insolubility of $Fe(OH)_3$, very little iron Fe^{3+} will be dissolved in natural waters when the pH is in the 7.0-8.5 range. Most of the natural waters that contain high levels of iron are groundwater with high concentrations of the fairly soluble Fe^{2+}. Since groundwater is protected from exposure to oxygen, any Fe^{2+} that dissolves from surrounding minerals is not oxidized to Fe^{3+}. Because Fe^{2+} is so much more soluble than Fe^{3+}, more iron can remain in solution.

What do you already know?

As you learned in Exploration 2A, the key feature of any reliable analytical technique is that the analysis be *specific* for the analyte (the substance of interest), even in the presence of other substances. By the end of this Exploration, you will have learned the answer to the question *How do you determine how much iron is in a water supply?*

Preparing for Inquiry

BACKGROUND READING

Determination of Iron in Water

The method used for iron determination makes use of **coordination chemistry**, or the ability of transition metals like iron to bond to a substance that has an unshared pair of electrons available. In many bonds, each atom contributes one electron. However, transition metals can act as **Lewis acids** (electron pair acceptors) and tend to form bonds with compounds that are **Lewis bases** (electron pair donors), species that contain an unshared pair of electrons. Examples of Lewis bases include water (H_2O), chloride (Cl^-), and ammonia (NH_3). These molecules or ions can donate *both* electrons of an unshared pair of electrons to an empty orbital on the metal. Groups that bind to metals in this fashion are called **ligands**.

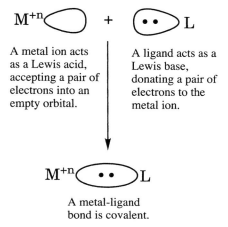

A metal ion acts as a Lewis acid, accepting a pair of electrons into an empty orbital.

A ligand acts as a Lewis base, donating a pair of electrons to the metal ion.

A metal-ligand bond is covalent.

Figure 2-4: Metal ions have empty orbitals that are available for sharing with ligands that have unshared pairs of electrons such as H_2O, Cl^-, and NH_3.

This type of bond is characteristic of the transition metals. In fact, it is rare to find a "bare" transition metal that does not have some other ligands bound to it because bare metal ions are quite unstable and very reactive. Clearly, it is energetically favorable for a positively charged ion to interact with a molecule or ion that has non-bonding electrons. For example, in aqueous solution, iron ions form bonds to six water molecules to make $[Fe(H_2O)_6]^{2+}$ or $[Fe(H_2O)_6]^{3+}$. The abbreviation for these species, $Fe^{2+}(aq)$ or $Fe^{3+}(aq)$, indicates that the iron ion is surrounded by water molecules, or **solvated**.

$$\left[\begin{array}{c} OH_2 \\ H_2O\cdots \underset{\displaystyle H_2O}{\overset{\displaystyle |}{\underset{\displaystyle |}{Fe}}} \cdots OH_2 \\ OH_2 \end{array} \right]^{+2}$$

One of the most fascinating aspects of coordination compounds is their varied and beautiful colors. The color usually depends on the atoms bound directly to the metal ion. For example, an aqueous solution of Fe(II) is a very pale yellow or green, but if certain nitrogen-containing compounds are added, a deeply colored purple solution results. We can use this property of coordinate bonds to measure the concentration of a metal ion in solution. If the sample is treated with an appropriate ligand that will transform it into a colored species, we can measure the intensity of the color and use this as an indication of how much iron is in solution. In general, the more concentrated the solution is in iron, the more intense the color. In order to obtain quantitative results for the amount of iron in solution, a technique called **spectrophotometric analysis** is used, described in more detail in the following section.

One reagent that coordinates to iron to form a colored complex is an organic molecule that has the trade name **ferrozine**. This molecule has two nitrogen atoms in a position to donate unshared electron pairs to the iron to form a purple iron-ferrozine complex (Figure 2-5).

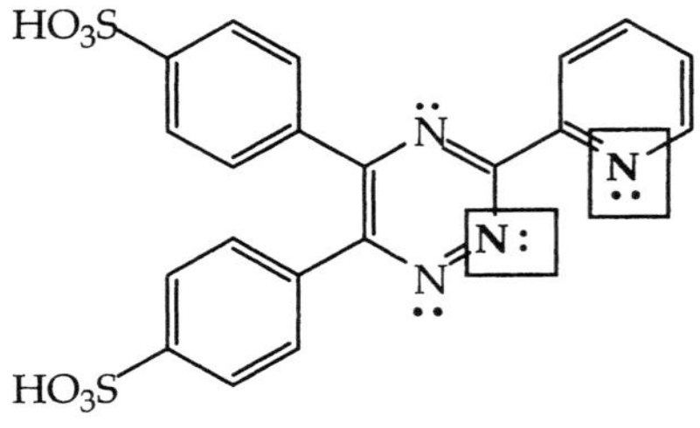

Figure 2-5:   *Ferrozine contains nitrogen atoms with unshared pairs of electrons that will bind to Fe(II) ions in solution. The two nitrogens that are boxed in the figure are the ones that bind to iron.*

Ferrozine does not react with iron (III), so in the ferrozine solution, a mild reducing agent, hydroxylamine (NH_2OH), is added to transform all iron (III) in solution to iron (II). A pH 5.5 buffer is also added to ensure that the solution stays within the optimum pH range for reaction of ferrozine reagent with iron (II). The ferrozine solution is concentrated enough to ensure that essentially all the iron (II) reacts to form the iron (II)-ferrozine complex (Figure 2-6).

purple iron-ferrozine complex

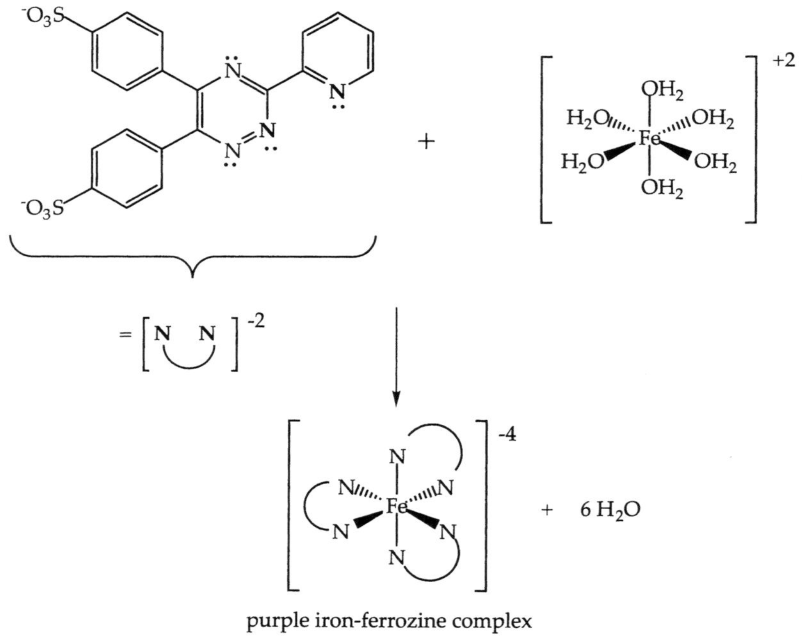

Figure 2-6:   *Ferrozine reacts with solvated iron(II) to form a deep purple iron (II)-ferrozine complex.*

Colorimetric Methods of Analysis

As you learned in the previous section, when ferrozine is added to a solution containing iron (II), an iron (II)-ferrozine complex forms and the solution turns a deep purple color. Because the intensity of this color is related to the amount of iron (II) present, we can measure the concentration of iron (II) in different solutions based on their color. Chemists call this analytical technique **colorimetric analysis**. Colorimetric analyses can be used to measure the concentration of many different constituents in water by reacting the chemical species of interest with a specific reagent that produces a colored compound in solution.

Colorimetric analysis is carried out using a **spectrophotometer**, an instrument that measures the amount of light transmitted through a solution (Figure 2-7).

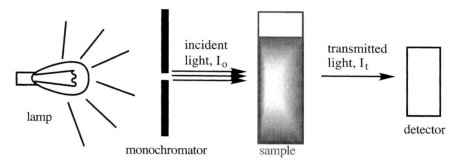

Figure 2-7: The spectrophotometer consists of a visible light source with a monochromator that allows the analyst to select a small range of wavelengths for the analysis. Light passing through the sample is absorbed by the species of interest and the intensity of the transmitted light (I_t) is detected.

When light is passed through any solution, some of the light is absorbed or reflected, and the remainder is transmitted through the solution. The amount of light transmitted through a solution depends on many factors, including the wavelength of the incident light, the color of the solution (and hence the concentration of the substance of interest), and the path length of the light.

For a simple analogy, consider how a flashlight beam might behave if you were to shine it through a glass of water. Clean water does not significantly absorb or scatter light, and you perceive the light beam to be nearly as intense after passing through the water as it was before. Now imagine adding a small amount of blue food coloring to the water. The solution would turn blue, and the same beam of light transmitted through the solution would be dimmer than the light coming out of the flashlight itself. The **transmittance**, T, of a solution is the ratio of the intensity of the transmitted light, I_t, to the incident light, I_o:

$$T = \frac{I_t}{I_o}$$

The Beer-Lambert Law

A spectrophotometer can be used to measure either the transmittance or the **absorbance** (A) of a solution. As its name implies, absorbance is a measure of the amount of light absorbed by a solution. The relationship between absorbance and transmittance is logarithmic:

$$A = -\log T$$

The absorbance of a solution is directly proportional to the concentration of the substance of interest. The relationship between absorbance and concentration is given by the Beer-Lambert Law:

$$A = \varepsilon bc$$

where A is the absorbance (a unitless number determined by the spectrophotometer), b is the path length of the light through the sample (typically 1 centimeter for a standard spectrophotometer), c is the concentration of the substance of interest in solution, and ε (the Greek letter epsilon) is a constant that is specific for the substance being analyzed. In practice, ε is determined experimentally for each chemical species.

In general, the higher the concentration of the substance of interest in solution, the greater the absorbance measured by the spectrophotometer. Before the concentration of a substance of interest can be measured using a spectrophotometer, a plot of absorbance versus concentration, also known as a **standard curve**, must be constructed for the chemical species (see Figure 2-8). The slope of the resulting line is εb. Using this slope value and the Beer-Lambert Law, the concentration of the species of interest in any solution can be determined simply by measuring its absorbance.

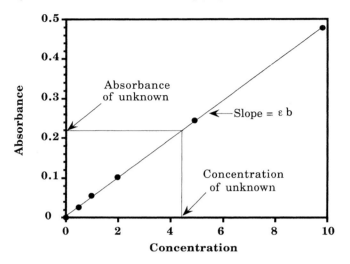

Figure 2-8:   The standard curve describes the linear relationship between absorbance and concentration: $A = \varepsilon bc$.

PRELAB QUESTIONS

Read the **Background Reading** section on Iron on pages 36–40 and the **Procedures** for determining iron on pages 40–42. Answer the following questions in your laboratory notebook before arriving in the laboratory.

1.   What chemical compounds are the source of iron in natural waters? With what minerals are they associated?

2.   What problems are associated with too much iron in a water supply?

3.   Write the chemical reaction used in the analysis of iron, drawing the structures of the products and reactants.

4.   What values will you graph after using the spectrometer to measure the absorbance of the standard solutions?

5.   Briefly outline the procedure you will follow to determine the concentration of iron in your sample.

Developing Ideas

PROCEDURES FOR DETERMINING IRON

In this experiment, you will determine the concentration of iron in your sample by a colorimetric method. Background information on the technique can be found on pages 36-40.

---

note

All glassware used in this experiment should be very clean and well rinsed with distilled water. If the glassware will not come clean with normal procedures, use the procedures in Appendix 2A.

Calibrating the Spectrophotometer

The spectrophotometer must be calibrated to correlate the absorbed light from a sample to a particular iron concentration. Thus, you must begin by running four reference standards with known iron concentrations: 0.1 ppm, 0.5 ppm, 1.0 ppm, and 3.0 ppm. In addition, the analysis also requires a blank, or a solution with all of the reagents added *except* iron. Water and other reagents absorb a small amount of light, and this blank allows you to measure that amount so that you can subtract it from the light absorbed by the purple iron-ferrozine compound. For more information about standard curves, see p. 40.

---

note

The wavelength of light at which you will measure the absorbance of the solution is 562 nm. Remember to set your spectrophotometer to the correct setting!

---

1. Prepare a table in your lab notebook with Sample ID and concentration in one column and absorbance in another. Use this table to record your measurements.
2. Label a set of five test tubes or special cuvettes with a number corresponding to the concentration of each standard and the blank sample. Fill the cuvettes with the appropriate sample and note the color intensity of each in your notebook. What trends do you observe?
3. Wipe the outside of the blank cuvette with a Kimwipe™ and place it in the instrument. Follow the instructions for using your spectrophotometer to first zero the instrument with the blank solution.
4. Using the same procedures, place each of the standards in the spectrometer and read the Absorbance. Enter the data in the table in your notebook.
5. Construct a standard curve by plotting Absorbance (*y*-value) as a function of concentration (*x*-value). Draw the best straight line through the data points and calculate the slope of the line (see Figure 2-8).

At this point, you are ready to analyze samples.

---

note

The spectrometer needs to be calibrated on a regular basis, at least once every hour. For the most reliable data, calibrate the spectrometer *every time you use it*.

Preparing the Sample for Iron Analysis

1. Each member of your team should carry out one analysis of the original impure water sample. As you work on removing the contaminant or **remediating** the sample (see Session 6), you will use the same analytical procedure. For the analysis of the remediated sample, a *single* analysis by one team member will be sufficient.
2. Begin by obtaining several cuvettes or test tubes. Be sure they are clean and dry before use.

3. Using a 10.0 mL graduated cylinder, measure out 3.0 mL of your water sample, and add it to a cuvette. Make this measurement as accurately as possible!

4. Using a Re-pipet or a very accurate pipet, add 0.1 mL (2 drops) of the ferrozine reagent. Cap (with parafilm, if available) the cuvette or test tube and invert to mix the solution. Allow to stand for at least 15-30 seconds.

5. Using a Re-pipet or a very accurate pipet, add 0.1 mL (2 drops) of pH 5.5 buffer solution to the cuvette. Cap the cuvette or test tubes and invert to mix the solution. Wait at least two minutes for the color to develop.

6. Repeat the procedure for each replicate sample, then proceed as follows.

Determining Iron in a Sample

1. Prepare the sample for analysis as above. Zero the (calibrated!) spectrophotometer using the blank, then place the sample cuvette into the spectrophotometer and follow the instructions for your spectrophotometer to obtain an Absorbance reading. Record this value in your laboratory notebook.

2. For more precise measurements (original sample and final treated sample), analyze at least three replicate samples.

Calculations

Use the Beer-Lambert Law (see page 39) and the slope of your standard curve to calculate the concentration of iron in your sample. Average the values from your replicate analyses. The analyses should differ by less than 5%. If this is not the case, repeat the sample preparation and analysis if time permits.

Dispose of all chemical waste in the container provided.

disposal

Working with the Ideas

The following problems will help you understand the issues in more depth.

1. What is the maximum allowable contaminant level (MCL) for iron?

2. In your own words, describe the theory behind the spectroscopic method used for measuring iron.

3. A student in a different class comes into the lab to use the spectrophotometer for a phosphate analysis that produces a yellow solution. When he places his highest standard in the spectrometer, the absorbance reading is nearly zero. What could be wrong with his procedure? List as many reasons as you can.

Exploration 2E

How do you determine the total alkalinity in a water supply?

Creating the Context

Why is this an important question?

Many water supplies naturally contain mineral salts composed of ionic species that can neutralize acid and base and thereby **buffer** the system against a change in pH (see Session 9 for more information on buffer solutions). The capacity of a solution to neutralize acid is called **acid-neutralizing capacity** (**ANC**) or **total alkalinity**. The compounds that are responsible for total alkalinity in natural waters are typically mineral salts containing carbonate (CO_3^{2-}), bicarbonate (HCO_3^-), and hydroxide (OH^-) anions, with calcium (Ca^{2+}) and magnesium (Mg^{2+}) as the positively charged counterions (see Table 2-2).

Table 2-2   Minerals with Acid-Neutralizing Capacity

Mineral	Chemical Composition
Calcite	$CaCO_3$
Dolomite	$CaCO_3 \cdot MgCO_3$
Magnesite	$MgCO_3$ (rare)
Brucite	$Mg(OH)_2$ (rare)

Water that is too acidic will dissolve toxic metal ions from pipes. Thus, some degree of alkalinity is desirable so the water does not become too acidic in the distribution pipes. The presence of alkalinity reduces the amount of water treatment chemicals that must be added to remove other contaminants from the water as well.

However, because the alkalinity ions are accompanied by calcium and magnesium (the water hardness ions), water that is too high in alkalinity can be problematic. Insoluble precipitates tend to form when waters that contain high concentrations of calcium and magnesium and their negatively charged counterions carbonate (CO_3^{2-}) and bicarbonate (HCO_3^-) are heated. These precipitates, known as "scale," can eventually clog water delivery pipes, reducing the efficiency of water flow.

What do you already know?

As you learned in Exploration 2A, the key feature of any reliable analytical technique is that the analysis be *specific* for the analyte (the substance of interest), even in the presence of other substances. The total alkalinity analysis is specific for the carbonate (CO_3^{2-}), bicarbonate (HCO_3^-), and hydroxide (OH^-) ions in water. By the end of this Exploration, you will have learned the answer to the question *How do you determine the total alkalinity in a water supply?*

Preparing for Inquiry

BACKGROUND READING

A Brief Introduction to Acids and Bases

To understand the chemistry behind the total alkalinity measurement, we will need to learn a little bit about **acids** and **bases**. Later, in Sessions 7 and 8, we'll learn more.

The **Brønsted-Lowry** definition characterizes an **acid** as a substance that can act as a proton (H^+) donor. In the following equation, hydrochloric acid, HCl, is acting as an acid, donating a proton to water.

$$HCl\,(aq)\ +\ H_2O\,(l)\ \longrightarrow\ H_3O^+\,(aq)\ +\ Cl^-\,(aq)$$

A **base** is a substance that can act as a proton acceptor. In the following equation, aqueous ammonia is acting as a base, accepting a proton from water. Note that water can act as either an acid or a base.

$$NH_3\,(aq)\ +\ H_2O\,(l)\ \longrightarrow\ NH_4^+\,(aq)\ +\ HO^-\,(aq)$$

An acidic solution contains more **hydronium ions (H_3O^+)** than **hydroxide ions (OH^-)**, and a basic solution contains more hydroxide ions than hydronium ions. The **pH scale** is used to describe the concentration of acid or base in a solution, covering a range of ~ 0-14 pH units. A neutral solution at 25°C has a pH of 7.0; a basic solution has a pH of more than 7.0; an acidic solution has a pH of less than 7.0.

The alkalinity ions CO_3^{2-}, HCO_3^-, and OH^- are all bases, and as such, react readily with acid to form water and the protonated form of the ion, thus reducing the concentration of acid in water, as follows.

Carbonate:

$$H_3O^+(aq)\ +\ CO_3^{-2}\,(aq)\ \longrightarrow\ HCO_3^-\,(aq)\ +\ H_2O\,(l)$$

Bicarbonate:

$$H_3O^+(aq)\ +\ HCO_3^-\,(aq)\ \longrightarrow\ H_2CO_3\,(aq)\ +\ H_2O\,(l)$$

Hydroxide:

$$H_3O^+(aq)\ +\ HO^-\,(aq)\ \longrightarrow\ 2\,H_2O\,(l)$$

Reactions of this type are called **neutralization** reactions. We will take advantage of the reactions just shown for the analysis of total alkalinity. Because we know the chemical equations describing these reactions and therefore the stoichiometry of the reacting species, we can determine the alkalinity of a sample by determining how much acid can be added without making the sample acidic.

The key to the analysis is to have some means of determining when the solution has become acidic and has therefore used up all of the alkalinity present. Colored acid-base **indicators** are most commonly used for this purpose. Indicators are large organic molecules that are acids or bases themselves. Depending on the pH of the solution, they will adopt a specific color. The defined endpoint for the total alkalinity titration is a pH of 4.5, and bromocresol green is the indicator used for the titration. While this pH is far below neutral, we will see in Session 9 that a pH of 4.5 best describes the situation when all of the carbonate and bicarbonate anions in a solution are used up. Since these two anions are the predominant alkalinity ions in natural waters, the endpoint has been defined to reflect this reality.

Determination of Total Alkalinity

Experimentally, total alkalinity is determined by titrating a known volume of sample with acid, usually H_2SO_4 or HCl, in the presence of bromocresol green indicator. For example, titration of the carbonate ions resulting from dissolution of calcium carbonate can be described by the following neutralization reactions:

$$CaCO_3 \text{ (aq)} + H_2SO_4 \text{ (aq)} \longrightarrow H_2CO_3 \text{ (aq)} + CaSO_4 \text{ (aq)}$$

Excluding spectator ions:

$$CO_3^{-2} \text{ (aq)} + 2\,H_3O^+ \text{ (aq)} \longrightarrow H_2CO_3 \text{ (aq)} + 2\,H_2O \text{ (l)}$$

The indicator bromocresol green is blue above pH 5.4 and yellow below pH 3.8. At the endpoint pH of 4.5, the solution will be green.

The reporting of total alkalinity is usually simplified by assuming all acid neutralizing capacity is due to the presence of $CaCO_3$, and is thus reported as **mg of $CaCO_3$ per liter**. (Alkalinity is frequently caused by substances other than $CaCO_3$, so this reporting procedure is not precise. However, the net effect in terms of acid neutralizing capacity is essentially the same.)

Expected Total Alkalinity Values

The range of total alkalinity values for natural waters is typically between 30 and 500 mg of $CaCO_3$/L, with higher values occurring in regions that have alkaline soils. Rainwater usually has very little total alkalinity (<10 mg/L) since it has contact with few minerals. Surface waters generally have total alkalinities less than 200 mg/L, while groundwater total alkalinities are frequently much higher, sometimes over 1,000 mg/L due to a higher partial pressure of CO_2 in the subsurface from microbial degradation of organic matter underground. The underground CO_2 reacts with water to form bicarbonate and acid.

$$CO_2 \text{ (g)} + 2\,H_2O \text{ (l)} \longrightarrow HCO_3^- \text{ (aq)} + H_3O^+ \text{ (aq)}$$

The acid (H_3O^+) formed in this reaction reacts with calcium or magnesium carbonate in the surrounding rock to dissolve these compounds and produce bicarbonate, which is the main contributor to total alkalinity.

$$H_3O^+ \text{(aq)} + CaCO_3 \text{ (s)} \longrightarrow H_2O \text{ (l)} + HCO_3^- \text{ (aq)} + Ca^{+2} \text{ (aq)}$$

PRELAB QUESTIONS

Read the **Background Reading** section on total alkalinity on pages 43–45 and the **Procedures** for determining total alkalinity on pages 46–47. Answer the following questions in your laboratory notebook before arriving in the laboratory.

1. What chemical compounds are the source of total alkalinity in natural waters? With what minerals are they associated?

2. What problems are associated with too much alkalinity in a water supply? With too little alkalinity in a water supply?

3. Consider the carbonate ion, CO_3^{2-}, a major contributor to alkalinity in natural waters. Is this ion a Brønsted acid or a Brønsted base? Explain, using a chemical equation to show the acid-base reaction.

4. Why is total alkalinity also referred to as "acid neutralizing capacity"? Write the relevant chemical equations that support your answer.

5.   Briefly outline the procedure you will follow to determine total alkalinity in the laboratory.

Developing Ideas

PROCEDURES FOR DETERMINING TOTAL ALKALINITY

In this experiment, you will determine the concentration of total alkalinity in your sample by a titrating the sample with acid in the presence of an acid-base indicator. Background information on the technique can be found on pages 43–45.

---

note   All glassware used in this experiment should be very clean and well rinsed with distilled water. If the glassware will not come clean with normal procedures, use the procedures in Appendix 2A.

---

1.   Prepare the sample for the total alkalinity determination by measuring 20 mL of sample (measured with a graduated cylinder) into a clean 250 mL Erlenmeyer flask. Add 25 mL of distilled water and a drop or two of bromocresol green indicator. If the pH of your sample is below 4.5 (indicator is yellow), you do not need to determine alkalinity, since the sample will have *no* acid neutralizing capacity. If the pH is *above* 4.5, proceed to the next step.

2.   Prepare your buret by filling it to the top of the graduated area with standard 0.0100 M sulfuric acid, H_2SO_4.

---

note   The concentration of the acid prepared for your class will NOT be exactly 0.0100 M. Be sure to write down the EXACT concentration from the bottle you used to fill your buret.

---

3.   Check that the buret tip contains no air bubbles. If it does, allow a few mL of acid to flow out. You may need to tap the buret to dislodge the air bubbles.

4.   Take the initial volume reading from the buret. Always have your eyes at the same height as the liquid and read the value at the bottom of the meniscus to one decimal point (e.g., 10.1) and estimate the second decimal point.

5.   Swirling the sample gently, titrate the solution with 0.0100 M sulfuric acid (H_2SO_4) to the pH 4.5 endpoint, where the indicator changes color from blue to green. If you pass the endpoint by more than a drop or two, the solution will be yellow and you should redo the titration.

Calculations

1.   From the volume of acid used to titrate the sample, calculate the total alkalinity in millimoles of $CaCO_3$ per liter. This is a four-step process:
   •   Use the concentration of H_2SO_4 to convert mL of H_2SO_4 used in the titration to *moles* of H_2SO_4
   •   Use the mole ratio of reactants in the reaction of $CaCO_3$ and H_2SO_4 to determine the moles of $CaCO_3$ that must have been present initially.
   •   Use the volume of the sample to obtain moles of $CaCO_3$ per liter of sample.
   •   Convert moles per liter to millimoles per liter and report your concentration of alkalinity in millimoles per liter.

2.   The convention in water chemistry is to treat alkalinity as if it were *all* due to $CaCO_3$. To report your alkalinity values in a way that can be compared to the

water quality standards, convert your alkalinity in millimoles per liter to total alkalinity in mg of $CaCO_3$ per liter.

Working with the Ideas

The following problems will help you understand the issues in more depth.

1.   In your own words, describe the theory behind the titrimetric method used for measuring total alkalinity.

2.   If the only source of alkalinity in a sample is sodium hydroxide (NaOH), how would the calculation for total alkalinity be different?

Exploration 2F

How do you determine the concentration of Total Dissolved Solids (TDS) in a water supply?

Creating the Context

Why is this an important question?

What do you already know?

Water quality engineers often use more general analytical techniques than the specific ones you have just encountered for fluoride, water hardness, and iron. The measurement of total dissolved solids (TDS), for example, is a general but fast and accurate estimate of water quality.

As you learned in Exploration 1D, you can obtain a measurement of the total concentration of just the *ionic* components in solution by taking advantage of the fact that ions are charged species. A solution containing charged species will act as a conductor and permit the flow of electricity through the solution, with the amount of current flow proportional to the concentration and identity of dissolved ions in solution. The total dissolved solids measurement provides a way to estimate the overall concentration of dissolved ionic substances in water. The technique can be used in the field to track sources of pollutants, such as runoff from a farmer's feedlot, discharges by industry, leachate from mining wastes, or wastewater discharge.

As you learned in Exploration 2A, the key feature of any reliable analytical technique is that the analysis be *specific* for the analyte (the substance of interest), even in the presence of other substances. The total dissolved solids analysis is specific for all ionic substances and is a good general measure of water quality. By the end of this Exploration, you will have learned the answer to the question *How do you determine the concentration of total dissolved solids (TDS) in a water supply?*

Preparing for Inquiry

BACKGROUND READING

A solution of pure water has a very high resistance to the flow of electricity. The electrical **resistance** of the water is lowered by dissolution of an ionic compound in the solution. The ions provide a pathway for the flow of charge through the solution. The **conductance** of a solution is simply the reciprocal of the resistance and is a measure of the solution's ability to conduct electricity. The unit of measure for conductance is the **Siemen (S)**. Conductances in freshwater systems are usually on the order of microSiemens (μS), i.e., 10^{-6} Siemens.

In the laboratory, the instrument used to measure total dissolved solids is a **conductivity meter**, a sensitive electrical circuit called a Wheatstone bridge that measures the conductance of the solution between two parallel plates submerged in the sample. **Conductivity** is the quantity that is usually recorded as a measure of total dissolved solids, where conductivity is conductance per unit distance between the two parallel plates, usually expressed as Siemens per centimeter (S/cm) or, for smaller conductivities, microSiemens per centimeter (μS/cm). The conductivities of some common water sources are given in Table 2-3.

Table 2-3   Representative Conductivities

Water Source	Conductivity (μS/cm)	Source of Conductivity
Deionized water	< 1	----
Rainwater	10-40	NO_3^-, SO_4^{2-}, H_3O^+
Unpolluted surface waters	30-400 (higher in alkaline, desert areas)	Ca^{2+}, Mg^{2+}, Na^+, K^+, HCO_3^-, SO_4^{2-}, Cl^-, NO_3^-
Wastewater treatment plant effluent	300-1000	Ca^{2+}, Mg^{2+}, Na^+, K^+, HCO_3^-, SO_4^{2-}, Cl^-, NO_3^-, PO_4^{3-}

Another way to determine the total dissolved solids is the gravimetric method. You take a known volume of the sample of water, filter it, and carefully evaporate the water. When all the water has evaporated, the constituents that were dissolved in the water form a dry residue. You then weigh this residue to determine the weight of dissolved solids in mg per liter of water. This method is more time-intensive than a conductivity measurement, but the equipment is less expensive.

PRELAB QUESTIONS

Read the **Background Reading** section on Total Dissolved Solids (TDS) on pages 48–49 and the **Procedures** for determining TDS on pages 49–51. Answer the questions below in your laboratory notebook before arriving in the laboratory.

1. What general class of chemical compounds is the source of TDS in natural waters?

2. List two methods of determining TDS in water.

3. Briefly outline the procedure you will follow to determine the concentration of TDS in your sample.

Developing Ideas

PROCEDURES FOR DETERMINING TOTAL DISSOLVED SOLIDS (TDS)

In this experiment, you will determine the concentration of total dissolved solids in your sample either by measuring the conductivity of the solution *or* by a gravimetric determination. Background information on the techniques can be found on pages 48-49.

note   All glassware used in this experiment should be very clean and well rinsed with distilled water. If the glassware will not come clean with normal procedures, use the procedures in Appendix 2A.

Measuring TDS Using Conductivity

1. Rinse the conductivity cell with deionized water and shake it gently to remove the wash water. Place the cell into the sample, making sure that it is covered

completely with solution. Agitate it gently to remove air bubbles. Measure the temperature of the solution with a thermometer and record the temperature in your laboratory notebook.

2. Turn the meter on. If your meter is of the Wheatstone bridge type, balance the cell to obtain a null reading on the meter. On other meters, the value can simply be read from the display. Check with your instructor for specific instructions for using your instrument. In your notebook, record the value for the conductivity of the solution in microsiemens per centimeter (μS/cm).

---

 note   The term *mho* is an older term that is the same as Siemen. The two are interchangeable.

---

3. If the temperature of the solution is not 25°C, correct the reading using the following equation.

$$C_{25} = C_t (1 - 0.025\ \Delta t)$$

where C_{25} is the conductivity at 25°C, C_t is the conductivity measured at the sample temperature, and Δt is the difference between the sample temperature and 25°:

$$\Delta t = t\ (\text{sample}) - 25°C$$

If the temperature is less than 25°C, Δt is negative. If the temperature is greater than 25°C, Δt is positive.

4. Turn the meter off. Thoroughly rinse the cell with deionized water and place it in a beaker of clean deionized water, ready for the next person to use.

 At this point, you have not added any hazardous substances to the sample, so it may be disposed of down the drain.

disposal

Measuring TDS Using a Gravimetric Method

The concentration of total dissolved solids in a solution can be determined by filtering the sample, evaporating the water at 103–105°C, and weighing the residue remaining after the water evaporates. The concentration of total dissolved solids determined by this method is given in milligrams per liter and includes not only ionic substances, but also non-volatile organic substances that are water soluble.

1. Label a clean 150-mL beaker with your sample number and place it in a drying oven at 103–105°C for at least 1 hour. Remove the beaker from the oven, place it in a desiccator until cool, then weigh the beaker to the nearest 0.1 mg, recording the weight in your laboratory notebook.

2. Filter 50.0 mL of sample through 0.45 μm filter paper into the clean, dry beaker.

3. Place the beaker on wire gauze over a Bunsen burner and heat the sample to just below boiling to reduce the volume of the sample to approximately 10 mL. Do not allow the sample to boil over or splatter.

4. Allow the beaker to cool and place it in the oven at 103–105°C overnight or until the next laboratory period.

5. Remove the beaker from the oven and place it in a desiccator to cool to room temperature. Reweigh it to the nearest 0.1 mg, recording the weight in your laboratory notebook.

Calculations

Calculate the total dissolved solids in milligrams of solids per liter of solution. To convert conductivity in μS/cm to TDS in mg/L, divide by a factor of 1.1. This factor allows you to estimate the amount of solids in mg/L by first assuming a typical ratio of ions in natural waters and then taking into account the ability of each of these ions to conduct electricity in water.

Working with the Ideas

The following problems will help you understand the issues in more depth.

1. What is the maximum allowable contaminant level (MCL) for TDS?
2. In your own words, describe the theory behind the conductivity method used for measuring TDS.
3. During the evaporation step in the gravimetric method of measuring TDS, some of the bicarbonate (HCO_3^-) in the sample is transformed into CO_3^{2-} and $CO_2(g)$. How will this transformation affect the accuracy of the TDS measurement?
4. What does a TDS measurement *not* tell you? If the TDS exceeds the MCL, what other testing steps should you take?

Making the Link

Looking Back: What have you learned?

Chemical Principles

What chemistry experience have you gained?

In the process of working through Session 2, you have gained experience in working with some important principles and techniques of chemistry. An understanding of these principles and techniques is valuable in solving a wide range of real-world problems. Your experience with Session 2 should have developed your skills to do the items on the following lists.

Laboratory measurements

You should be able to:

- Determine the concentration of fluoride in an aqueous solution and explain how the fluoride-selective electrode works (**Exploration 2B**), or
- Determine the water hardness (Ca^{2+}/Mg^{2+} concentration) of an aqueous solution and explain how an EDTA titration works (**Exploration 2C**), or
- Determine the concentration of iron in an aqueous solution and explain how the ferrozine colorimetric analysis for iron works (**Exploration 2D**).
- Determine the total alkalinity of an aqueous solution and explain how an acid-base titration works (**Exploration 2E**).
- Determine the concentration of total dissolved solids in an aqueous solution and explain how the electrochemical method of analysis for TDS works (**Exploration 2F**).

- In addition to knowing the details of the technique you performed, you should also be familiar with the other two analyses and the theories behind them.

More generally, you should be able to describe the following analytical methods:

1. **Titrimetric:** A titrimetric method relies on a chemical reaction that occurs when a reagent that reacts specifically and quantitatively with the **analyte**, or species of interest, is added dropwise to a sample until all of the analyte has been used up. The amount of reagent is stoichiometrically related to the analyte and can be calculated using volume and concentration information. Titrimetric reactions require some indication of when the reaction is complete; visual changes such as the color change of an indicator, precipitation, or dissolution are commonly used (**Explorations 2C** and **2E**).

2. **Colorimetric:** Colorimetric methods measure the concentrations of colored substances in solution. In general, the darker the color, the more concentrated the substance. Some constituents in water supplies are colored, but more often an excess amount of a reagent is added that reacts specifically and quantitatively with the analyte to produce a colored compound. Spectroscopy is then used to measure how much of the colored compound is in solution. Standards are required to calibrate the analysis (**Exploration 2D**).

3. **Electrochemical:** Electrochemical methods use a specially constructed electrical circuit composed of a current-measuring device and electrodes that respond specifically to an analyte of interest. The current flowing through a solution is proportional to the concentration of the analyte. Standards are required to calibrate some electrochemical methods, but not all (**Exploration 2B** and **2F**).

Working with units

You should be able to:

- Use mass/mole/volume relationships to determine the results of a titration (**Explorations 2C** and **2E**).
- Convert between mg/L and mol/L.

Thinking Skills

What general skills are you building for your resume?

In the process of working through Session 2, you have also been developing some general problem-solving and scientific thinking skills that are valued by employers in a wide range of professions and in academia.

Data analysis skills

You are able to:

- Use replicate analyses and statistical methods to verify the reliability of your results (**Explorations 2B, 2C,** and **2D**).
- Use standards to ensure the accuracy of an analytical procedure (**Explorations 2B** and **2D**).
- Use graphical methods to determine the mathematical relationship between two variables (**Exploration 2B** and **Exploration 2D**).
    - Graph linear relationships (**Exploration 2D**).
    - Graph non-linear relationships in linear form (**Exploration 2B**).

Checking Your Progress

What progress have you
made toward answering
the Module Question?

Appendix 2A

Procedures for Cleaning Glassware

Laboratory glassware may accidentally contaminate your sample if it is not scrupulously clean. Use the following procedures to clean equipment for water analysis.

Cleaning the Sample Container, Sampling Utensils, and Labware

In preparation for the Water Treatment laboratory, clean the following glassware:

- several of your smallest beakers
- 10 mL graduated cylinder
- 50 mL graduated cylinder
- two 250 mL Erlenmeyer flasks
- buret

---

caution

The acids used for cleaning are concentrated and will cause burns if spilled on the skin or in the eyes. You must wear goggles at all times. If you spill acid on your skin, flush the affected area with copious amounts of water. Know where the closest eyewash, shower, and sink are located. If you spill acid on the lab bench, put a *small* amount of baking soda on the spill to neutralize the acid and then wipe up the excess baking soda. DO NOT leave unlabeled containers of acid on your laboratory benchtop.

---

1. Inspect the glassware. Does it look dirty? Is it stained with rust or something unidentifiable? Do you know what was in it last? If the item is stained, go to Step 2. If the item looks clean, go to Step 3. If you are in doubt, ask your instructor. If you *know* that *you* washed it earlier, you may not need to wash it again. If someone else *told* you that he/she washed it, wash it again. Don't trust anyone else to wash your glassware! You need to *know* that it is clean.

2. If the item is very dirty, it might be necessary to dip it in a No-Chromix™ bath, a mixture of concentrated sulfuric acid and ammonium persulfate (**DANGER!! This reagent is very corrosive. Wear GOGGLES (not just safety glasses) and gloves when using this reagent**). Any work with No-Chromix™ should be done with the assistance of your instructor. If you spill any on yourself, flush the area immediately with water. After dipping the item in the cleaning bath, rinse the glassware first in a bucket of rinse water, then take it to the sink for the remainder of the washing process. You may skip to Step 7 if you have washed your glassware with No-Chromix™ first.

3. For glassware into which you can fit a brush, scrub it with soap and water with a bottle brush. If your brush is old and looks dirty, get a new one from your instructor. Rinse with tap water.

4. Obtain 25 mL of 8 M nitric acid (HNO_3), and **carefully** carry it back to your lab station. Rinse the item with the acid, being sure to wet all inside surfaces. This serves to oxidize and solubilize any oxidizable materials. You do not have to fill the item to capacity with the wash or rinse solution. Between 5 and 25 mL

should be enough. If the glassware is not visibly dirty, you may reuse the acid wash solution on other items.

5. Rinse with tap water twice. DO NOT REUSE THE RINSE WATER!

6. Obtain 25 mL of 1.2 M hydrochloric acid (HCl), and **carefully** carry it back to your lab station. Rinse the item with the acid, again being sure to wet all inside surfaces. You do not have to fill the item to capacity with the wash or rinse solution. Between 5 and 25 mL should be enough. If the glassware is not visibly dirty, you may reuse the acid wash solution on other items.

7. Rinse with tap water at least three times. DO NOT REUSE THE RINSE WATER!

8. Do a final rinse with de-ionized water, at least three times. DO NOT REUSE THE RINSE WATER! Do not dry items with a towel, as this will contaminate them again. Do not touch the part of the item that will come in contact with the sample with your fingers, as they are quite "dirty" if you are doing analysis for trace levels of contaminants. Store the items upside down so no dust or airborne bacteria fall onto the surfaces that might come in contact with a sample.

Do not pour acids down the drain. Dispose of all used acid solutions in the waste containers provided.

disposal

Dissolution Reactions

Why *do substances dissolve in water?*

Exploration 3A: The Storyline

In Session 1, you learned that a variety of substances are commonly found dissolved in natural waters. Not all substances are soluble to the same extent, however. For example, a teaspoon of salt will dissolve completely in a quart of water, forming a homogeneous solution of saltwater with no traces of the solid remaining. In contrast, a teaspoon of oil will form an oily layer on top of the water, with very little of the oil dissolved in the water. Why? What characteristics of a substance affect its solubility in water? More important for a water treatment engineer, how can we use our knowledge of these characteristics to remove undesirable constituents from water?

We'll begin this Session by observing the solubility behavior of several substances, to see if we can find trends related to their structure and composition. Other Explorations in this Session will examine the process of dissolution in more detail, with the goal being to answer the question, *Why do substances dissolve in water?*

Developing Ideas

PART ONE: EXPERT GROUPS

Working in teams of three or four, you will observe the solubility characteristics of the substances in a particular set of compounds (Table 3-1) and become **Expert** on that set. Team 1 will evaluate the compounds in Set 1, Team 2 will evaluate the compounds in Set 2, and Team 3 will evaluate the compounds in Set 3. Depending on the size of your class, more than one team may evaluate each set of compounds.

Each team should answer the following questions for their set of substances.

1. Before observing the solubilities of your set of compounds, place the compounds into categories based on the following characteristics. You should end up with several groupings, and record them in your notebook.

   a. Group the compounds by the elements in the compound, with particular attention to their locations in the periodic table.

   b. For Sets 1 and 2, group the compounds by the relative electronegativities of the elements in the compounds (see Table 3-2). Recall that electronegativity is the ability of an atom within a molecule to attract electrons to itself.

   c. Group the compounds by whether or not the substance is ionic, and the charge on any ions that might be part of a compound.

   d. Group the compounds by the structure and shape of the ion or molecule. (Your instructor may provide this information for you.)

Table 3-1   Sets of Compounds to Be Examined

Set 1	Set 2	Set 3
LiCl	ethanol CH_3CH_2OH 	KNO_3
Na_2O	n-octane $CH_3(CH_2)_6CH_3$ 	$CuSO_4$
$FeCl_3$	methanol, CH_3OH	$CaCO_3$
KI	n-hexane $CH_3CH_2CH_2CH_2CH_2CH_3$ 	$Ba(NO_3)_2$
MgO	n-butanol $CH_3CH_2CH_2CH_2OH$ 	Na_2CO_3
ZnO	n-pentanol $CH_3CH_2CH_2CH_2CH_2OH$ 	K_2SO_4
NaF	sodium lauryl sulfate $Na^+\,{}^-OSO_3(CH_2)_{10}CH_3$ 	$CuCO_3$
CuS	glucose $CH_2OHCH(OH)CH(OH)CH(OH)CH(OH)CHO$ 	$Cu(NO_3)_2$
$CaCl_2$	corn oil, $CH_3(CH_2)_8CH{=}CH(CH_2)_7COOH$ 	$CaSO_4$

Table 3-2   Information About Ions and Atoms

Common Simple Ions	Common Complex Ions	Pauling Electronegativities			
chloride, Cl^-	nitrate, NO_3^-	F	4.0	Fe	1.8
bromide, Br^-	sulfate, SO_4^{2-}	O	3.5	Zn	1.6
iodide, I^-	carbonate, CO_3^{2-}	Cl	3.0	Mg	1.2
oxide, O^{2-}		Br	2.8	Li	1.0
sulfide, S^{2-}		S	2.6	Ca	1.0
hydroxide, OH^-		C	2.5	Cu	1.0
		I	2.5	Na	0.9
		H	2.1	K	0.8

2. Mix 0.1 g of the solid compounds or 3–5 drops of the liquid compounds with 3 mL of water, observing the dissolution processs closely. Shake well and classify the substances into one of three categories:

   • Dissolves completely, forming a homogeneous solution,

   • Does not appear to dissolve at all, or

   • Appears to dissolve to some extent, but not completely.

3. What is the relationship between the solubility classification of each compound and the categories of compounds you generated in Question **1**?

4. Propose a set of solubility rules based on what you observed for your assigned set of compounds. Be sure that *each person* on the team can explain the team's rules. Record your rules in your notebook.

5. List at least two other compounds you would like to have available to test your rules.

PART TWO: CONSULTANT GROUPS

Now that your team is **Expert** at defining the solubility trends of your set of compounds and has a working hypothesis, the next step is to form **Consultant** teams. A **Consultant** team will consist of one member from each **Expert** team. Each **Consultant** team should do the following:

6. Have each person on the team describe the trends he or she observed for the solubility of the set of compounds studied in **Part One** and present the **Expert** team's solubility rules to the **Consultant** team. Be sure that everyone on the team notes the solubility characteristics of the compounds in *each set* in their laboratory notebooks. You will need this information in a later assignment.

7. Create a new set of rules that encompasses the observations of each **Expert** team. Record the revised set of rules in your notebook.

8. Pick a spokesperson to report the new rules to the class.

Working with the Ideas

9. Using what you learned in the preceding exercise, predict the solubility of the following compounds in water. Explain your reasoning.

    a. Methyl amine, CH_3NH_2

    b. Strontium oxide, SrO

    c. Calcite (see Table 1-2)

    d. Quartz (see Table 1-2)

    e. Magnesite (see Table 1-2)

    f. n-Octanol, $CH_3CH_2CH_2CH_2CH_2CH_2CH_2CH_2OH$

    g. Sodium sulfide, Na_2S

    h. n-Propanol, $CH_3CH_2CH_2OH$

10. What challenges do dissolved impurities pose for water engineers? How do these challenges compare to those faced when working with *insoluble* impurities?

Looking Ahead

Several Exploration Questions can guide you to explore the Session Question in more depth. These will be used at the discretion of your instructor.

*   **Exploration 3B:** What do we mean by the word "dissolve"?

*   **Exploration 3C:** What characteristics of a substance affect its solubility in water?

Exploration 3B

What *do* we mean by the word "dissolve"?

Creating the Context

Why is this an important question?

What do you already know?

The process of dissolution is critical for the survival of all organisms. Only through dissolution can essential minerals be taken up from the soil by plants or absorbed from a suspension of food in your stomach. At the cellular level, dissolved substances, such as glucose, sodium, potassium, calcium, and magnesium, and dissolved gases, such as carbon dioxide and oxygen, are transported in and out of cells, making it possible for the cell to carry out respiration and reproduction. Minerals are the source of many essential elements; through the dissolution of minerals, these elements become available for living things to utilize.

As you saw in Session 1, the minerals in rocks and soils dissolve when water comes in contact with them. Anthropogenic pollutants such as paint thinner, pesticides, and drycleaning solvents will also dissolve in water to some extent. Thus, we need to consider the process of dissolution when determining the potability of a water supply. If the water contains too many dissolved substances, or if the dissolved substances are toxic, the water supply may not be usable. Ideally, we'd like to know what substances are present near the water supply and learn more about their solubility properties so we can predict which substances may be present in quantities of concern. This will help us judge the suitability of a water source for human consumption.

In this Exploration, you will study dissolution of ionic substances on a molecular scale. You will create a model to explain the solubility properties you observed in Exploration 3A, and learn to use this model to predict the solubilities of other substances in water and in different solvents. By the end of this Exploration, you should be able to answer the question: *What do we mean by the word "dissolve"?*

Preparing for Inquiry

BACKGROUND READING

Water as a Polar Molecule

Before we can talk about the dissolution of minerals in water, we need to know something about the solvent we are considering, namely water. Many other substances can also act as solvents, but the one we care most about for developing a pure drinking water supply is water, so we will focus our attention on it.

The water molecule consists of two hydrogen atoms bonded to a single oxygen atom. The water molecule may be visualized with a Lewis dot structure or an equivalent structure that replaces the dots with lines, as follows.

The oxygen atom contributes six valence electrons, and each hydrogen atom contributes one, to form the requisite octet of electrons. The four electron pairs, two in each of the O-H bonds and two as unshared pairs, are arranged in a tetrahedral geometry so that the water molecule is bent, rather than linear. The shape of the water molecule and the two unshared pairs of electrons help determine the properties of water.

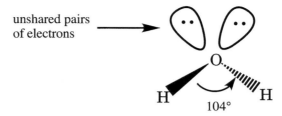

unshared pairs
of electrons

104°

Some atoms attract electrons more strongly than other atoms. Electronegativity (EN) refers to the strength with which an atom within a molecule attracts and retains electrons. The difference in electronegativity between hydrogen (EN = 2.1) and oxygen (EN = 3.5) results in an unequal sharing of electrons in the water molecule. This uneven distribution of charge is called **polarity.** The Greek symbol δ+ (delta plus) designates a partially positively charged area of the molecule and δ- (delta minus) designates a partially negatively charged area of the molecule. These symbols represent the *general* distribution of electrons in the molecule and are not quantitative. In this Exploration, you will determine which atom in the water molecule has a greater electron density. You will also examine the effects of this polarity on the dissolution of ionic substances.

Solutes in Water

The second factor to consider when thinking about dissolution is the species that is to be dissolved in water, the **solute.** In Exploration 3A, you looked at the solubility characteristics of two distinctly different types of compounds:

1. **Ionic compounds:** These species are composed of elements of very different electronegativities that combine through the force of attraction between positive and negative ions, the **ionic bond.** For example, NaCl is an ionic compound, with the sodium existing as an Na$^+$ ion and chlorine as a Cl$^-$ ion. Ionic solids are made up of an extensive repeating lattice of positive and negative ions (see Figure 3-1), so they cannot be labeled as individual, discrete molecules. Many minerals are ionic compounds, and most compounds made up of metals and non-metals are ionic compounds.

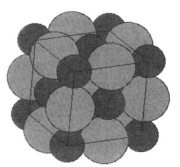

Figure 3-1:   Sodium and chloride ions form an infinite repeating lattice of positive and negative ions. In this drawing, the sodium ions are the darker, smaller spheres and the chloride ions are the lighter, larger spheres.

2. **Covalent compounds:** These species are composed of elements of similar electronegativities that combine by sharing electrons to form **covalent bonds.** Some examples are H$_2$O and the organic molecules ethanol, acetone (nail polish remover), and the mixture of compounds in gasoline. Most covalent compounds

Relative Bond Strengths (kJ/mol)
Ionic lattice   1,000-15,000
Single ion pair (g)   300-600
Covalent bond   150-500
Hydrogen bond   25-150
Intermolecular

exist as individual, discrete molecules. Collections of these molecules are held together by attractive **intermolecular forces**. These attractive forces exist because of charge differences in different parts of the molecules as a result of:

- **Permanent molecular dipoles:** differences in electronegativity between the component atoms, as seen for H_2O, or in chloromethane (Figure 3-2), or

- **Temporary induced dipoles:** temporary fluctuations in the positions of the electrons in a molecule composed of atoms of similar electronegativity. Intermolecular forces from temporary induced dipoles are called **London dispersion forces**.

These intermolecular forces can be viewed as very weak bonds (10-60 kJ/mol) that are continuously breaking and reforming at room temperature.

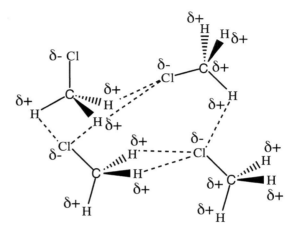

Figure 3-2:   Chloromethane molecules are held together by dipole-dipole interactions. These interactions serve as very weak bonds that are easily broken and reformed at room temperature.

The dissolution of both ionic and covalent compounds requires that the attractive forces between the ions or molecules in the solid state be disrupted by the solvent molecules. New bonds are then formed between the solvent molecules and the individual ions or molecules that comprise the solute.

A caveat: The labeling of a compound as ionic or covalent is not an either/or distinction, but a gradation from compounds that can best be described as oppositely charged spheres attracting each other (the ionic bond) to compounds that share electrons unequally (a polar covalent bond) to compounds that share electrons essentially equally (a non-polar covalent bond). In addition, many compounds, including a number of common minerals, consist of both ionic and covalently bonded units in the same mineral.

Developing Ideas

Your instructor will show you an animation of the dissolution of sodium chloride, which can also be found on the CD-ROM that accompanies this module. A smaller version is also on the ChemConnections web site at: http://chemistry.beloit.edu/Water/moviepages/Comp3salt.htm

The chemical equation that describes the dissolution is:

$$NaCl\ (s) + x\ H_2O\ (l) \longrightarrow Na^+\ (aq)\ +\ Cl^-\ (aq)$$

Working individually, write a description of the dissolution process, illustrated with sketches. In your description, you should address the following:

1.   Which atoms on the water molecules appear to be interacting most strongly with the *chloride* ion during the dissolution process?

2.   Which atoms on the water molecules appear to be interacting most strongly with the *sodium* ion during the dissolution process?

3.   Draw a picture of the water molecule that shows a distribution of charge consistent with what you observed when you answered questions **1** and **2**.

4.   Specifically, what bonds are being *broken* during the dissolution process?

5.   Specifically, what bonds are being *formed* during the dissolution process?

6.   When we write "$Na^+(aq)$" and "$Cl^-(aq)$", we indicate that sodium and chloride ions are dissolved in water. In your own words, describe what the words "dissolve" and "aq" mean from a microscopic point of view.

Working with the Ideas

The following problems will help you understand the concepts in more depth.

7.   You learned in Session 1 that the atmospheric gases oxygen and carbon dioxide dissolve in water to some extent. Draw the Lewis dot structure of each of these (covalent) gaseous molecules and then draw a picture of the interactions between water and these molecules that explains why they dissolve.

8.   Methanol, CH_3OH, is a covalent compound that is infinitely soluble in water.

Diagram the dissolution of methanol in water. Begin by showing a collection of 3 methanol molecules and 10 water molecules in separate containers, then sketch the mixture after the methanol has dissolved. Be sure to depict the shapes of the methanol and water molecules accurately. What bonds are broken? What bonds are formed?

9.   Apply the model you created for the dissolution of sodium chloride to the solubility tests you did in Exploration 3A. Can you explain all of your solubility rules based on this model?

Exploration 3C

What characteristics of a substance affect its solubility in water?

Why is this an important question?

When considering the safety of a drinking water supply, we need to determine what pollutants might be in the area and whether they will dissolve in water to a great enough extent to cause problems with potability. In this Exploration, we will examine the factors affecting the solubility properties of substances that might be found near a water supply. By the end of the Exploration, you should be able to answer the question: *What characteristics of a substance affect its solubility in water?*

As you watched the animation of the dissolution of sodium chloride in Exploration 3B, you probably noticed that only the oxygen atoms in the water molecule interacted with the sodium ion and only the hydrogen atoms of the water molecule interacted with the chloride ion. Thus, since opposite charges attract, the distribution of electrons in the water molecule must be such that the oxygen carries a slight excess of the electron density and the hydrogens are somewhat electron deficient. Because of these areas of separated charge, the entire water molecule has a net **dipole moment, μ,** defined as the product of the charge at either end of the dipole times the distance between the charges. The strength of a dipole moment is measured in units of **Debyes, D,** where 1 Debye = 3.33 x 10^{-30} Coulomb-meter and is represented by an arrow pointing towards the electron-rich, or negatively charged part of the molecule.

_____

What do you already know?

_____

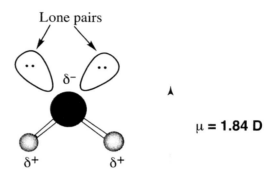

The oppositely charged dipoles of adjacent water molecules attract each other, resulting in the formation of molecular aggregates through intermolecular forces known as **hydrogen bonds**. These weak bonds form between hydrogens on one molecule and oxygens on another, resulting in a loosely bound network of water molecules. Hydrogen bonds are usually depicted as dotted lines connecting the hydrogen atom of one molecule to the oxygen atom of another. Figure 3-3 shows the intricate ring-shaped network of water molecules held together by strong intermolecular bonds. These molecular aggregates in water may be able to reach sizes of up to 100 water molecules at room temperature.

The strength of a hydrogen bond is on the order of 25–150 kJ/mole. By comparison, the strength of covalent bond ranges from 150 to 500 kJ/mole and the strength of a single ionic bond between two gaseous ions is on the order of 300–600 kJ/mole.

Although hydrogen bonds are not as strong as ionic or covalent bonds, they are strong enough to affect the properties of water significantly.

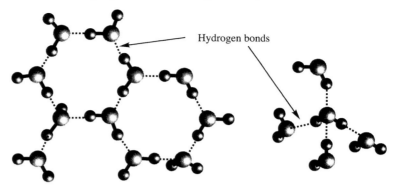

Hydrogen bonds

Figure 3-3: Water molecules form a strong network of intermolecular bonds between a hydrogen on one molecule and an oxygen on another. Left, a two-dimensional view of such a network. Right, a three-dimensional view, showing that each water molecule can form a total of four hydrogen bonds (dotted lines represent hydrogen bonds).

For water to dissolve a solute, the network of hydrogen bonds between water molecules must be disrupted. When dissolution occurs, individual hydrogen bonds between water molecules are broken and new bonds are formed between water molecules and the solute (see Figure 3-4). In general, *dissolution is favored when the energy released from forming all of these new bonds is greater than the energy required to disrupt the hydrogen bonds of the water and the interionic or intermolecular forces of the solute.* These bond energies represent the **enthalpy change** for the dissolution reaction, $\Delta \mathbf{H}_{solv}$.

Knowing the ΔH_{solv} can help you estimate the predicted solubility of a compound; however, for a thorough analysis of solubility, you also need to take into account the **entropy change**, $\Delta \mathbf{S}_{solv}$, of the dissolution reaction. The net **free energy change**, $\Delta \mathbf{G}_{solv}$, combines enthalpic and entropic effects and is the best predictor of solubility. For more information on how enthalpy and entropy affect the outcome of chemical reactions, see the chapters in your textbook on thermodynamics. In this Exploration, we will primarily examine the factors that influence the *enthalpy* changes for the dissolution reactions of molecular compounds such as organic molecules. This requires a closer look at the properties of water as a solvent.

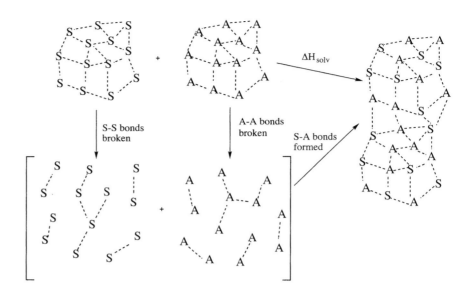

Figure 3-4: The dissolution of substance **A** in solvent **S** is favored when the energy required to separate particles of **S** and **A** is less than the energy given off by formation of **S-A** bonds, i.e., when ΔH_{solv} is negative.

Water is a particularly good solvent for ionic solutes because of its polarity, or uneven distribution of charge in the molecule. Water will surround a positively charged ion with the negatively charged part of the water molecule (or surround a negatively charged ion with the positively charged part of the water molecule), thereby isolating the ions from their neighbors and overcoming the attractive forces between positive and negative ions that maintain the crystal structure. This process is known as **solvation**; it is energetically favorable because of the electrostatic attraction between the ions and the charged areas of the water molecule (see Figure 3-5).

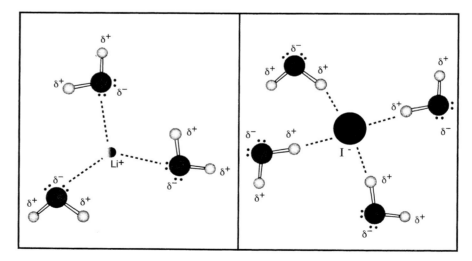

Figure 3-5: Water is a polar molecule and is therefore an excellent solvent for ionic compounds because it interacts strongly with charged ions. In this figure, the ionic compound lithium iodide is solvated by water.

The ion, stabilized by bonding to surrounding water molecules, is removed from the crystal lattice and carried into solution, becoming a solvated ion. This process is responsible for the dissolution of the minerals of the earth's crust. For similar reasons, water is also a good solvent for polar covalent molecules. For example, polar organic molecules like methanol (CH_3OH) dissolve readily in water, while non-polar organic molecules like pentane (C_5H_{12}) do not. You will explore the interactions between water and representative organic molecules in the Developing Ideas section of this Exploration.

A quantitative measure of the ability of a substance to separate charged particles is the **dielectric constant**, ε. Polar compounds with distinct regions of positive and negative charge have high dielectric constants. Conversely, non-polar compounds with very little charge difference across the molecule have low dielectric constants and are known as **non-polar** compounds. Water has a high dielectric constant and is one of the most polar solvents known (see Table 3-3).

Table 3-3 Dielectric Constants and Dipole Moments of Selected Liquids

Compound	Structure	Dielectric Constant (ε) at 25°C	Molecular Dipole Moment, μ (in Debyes, D)
Water		78.5	1.84
Methanol		32.6	1.70
Ethanol		24.3	1.69
Trichloromethane		4.8	1.01
Benzene		2.3	0.00

Developing Ideas

Work in teams of 3 or 4. Refer back to Table 3-1 on page 56 and examine the list of compounds in set 2, the organic compounds.

1. Classify each of these compounds as polar, non-polar, or amphipathic (containing both polar and non-polar parts).

2. Look back at the results of the solubility tests and label each one as soluble or insoluble in water. What is the relationship between the polarity of a compound and its water solubility?

3. Which of the compounds would you expect to have the strongest intermolecular forces, i.e., which molecule has the strongest attraction to other *like* molecules? Explain.

4. Which of the compounds would you expect to have the weakest intermolecular forces, i.e., which molecule would you expect to have the weakest attraction to other *like* molecules? Explain.

Working with the Ideas

The following problems will help you understand the concepts in more depth.

5. For each compound in Table 3-3 (page 67) with a dipole moment, draw an arrow to show the direction of the molecular dipole and put in partial charges (δ+, δ-) as in the picture of the water molecule on page 64.

6. Predict the solubility of all of the Set 2 compounds (shown below) in *hexane*, C_6H_{12}.

ethanol	n-pentanol
CH_3CH_2OH	$CH_3CH_2CH_2CH_2CH_2OH$
n-octane	sodium lauryl sulfate
$CH_3(CH_2)_6CH_3$	$Na^+ {}^-OSO_3(CH_2)_{10}CH_3$
methanol, CH_3OH	corn oil, $CH_3(CH_2)_8CH=CH(CH_2)_7COOH$ 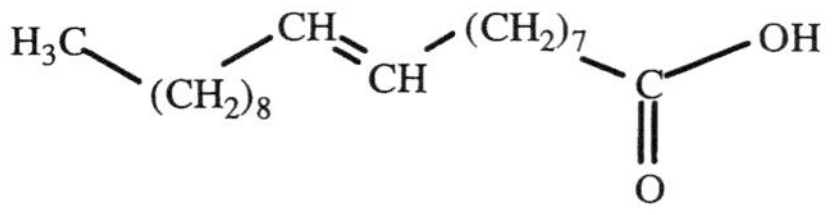
n-hexane	glucose
$CH_3CH_2CH_2CH_2CH_2CH_3$	$CH_2OHCH(OH)CH(OH)CH(OH)CH(OH)CHO$
n-butanol	
$CH_3CH_2CH_2CH_2OH$	

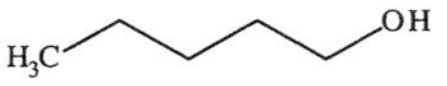

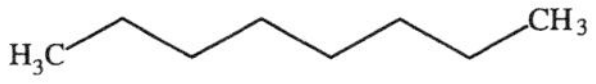

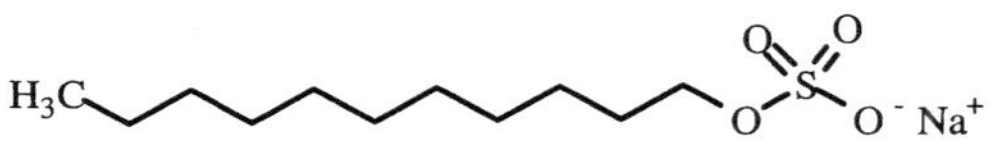

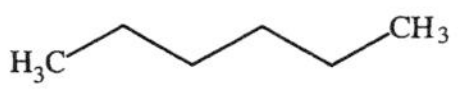

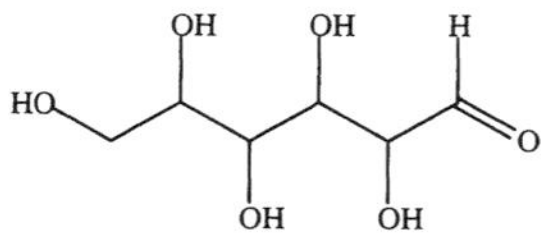

7. Three bottles labeled **A**, **B**, and **C** are filled with different liquids, one of either propanol, pentane, or 1-octanol. In the following table:

   a. label these compounds as polar or non-polar

    b. use the following information on solubility properties to identify A, B, and C.

- When 10 drops of compound **A** are added to several milliliters of water, a thin layer of liquid **A** is observed on top of the water.

- When 10 drops of compound **B** are added to several milliliters of water, the first few drops dissolve completely, but the end result is a thin layer of liquid **B** on top of the water.

- When 10 drops of compound **C** are added to several milliliters of water, it dissolves completely.

Compound	Polar or nonpolar?	A, B, or C?
Propanol, $CH_3CH_2CH_2OH$		
Pentane, $CH_3CH_2CH_2CH_2CH_3$		
1-octanol, $CH_3CH_2CH_2CH_2CH_2CH_2CH_2CH_2OH$		

8. The molecular strucures of three pesticides follow, along with their saturation concentrations in water in mg/L.

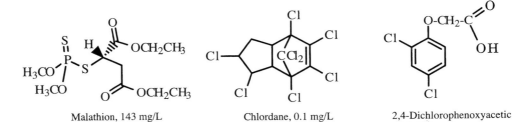

Malathion, 143 mg/L   Chlordane, 0.1 mg/L   2,4-Dichlorophenoxyacetic acid, 682 mg/L

    a. Using your knowledge of polarity and hydrogen bonding, explain the differences in the water solubility of these compounds.

    b. List several ways pesticides might make their way into a water supply.

    c. What do the solubilities tell you about the relative water-contaminating potential posed by each pesticide?

9. Organochlorine solvents, used for dry cleaning and grease removal in many industries, have polluted the groundwater supplies in several parts of the United States. Perchloroethylene, "perc," is one of these solvents.

$$Cl_2C=CCl_2$$

perchloroethylene

    a. Does perchloroethylene have a net molecular dipole moment? Explain, using the individual bond dipoles.

    b. Would you expect perchloroethylene to dissolve in water to a great extent? Explain.

10. You often hear people say that water is a universal solvent. Based on your discoveries in this Exploration, would you agree or disagree with this statement? Explain your answer.

Making the Link

Looking Back: What have you learned?

Chemical Principles

What chemistry experience have you gained?

As you worked through Session 3, you have gained experience with some important principles of chemistry. An understanding of these principles is valuable in solving a wide range of real-world problems. Your experience with Session 3 should have developed your skills to do the items on the following lists.

General Solubility

You should be familiar with:

- The solubility properties of some common anion/cation combinations (**Exploration 3A**).
- The difference between ionic and covalent compounds (**Exploration 3B**).
- The types of intermolecular forces and their relative magnitudes (**Explorations 3B and 3C**).
- Hydrogen bonding between water molecules (**Exploration 3B**).

The Dissolution Process

You should be able to:

- Draw a microscopic view of the process of dissolution (**Explorations 3B and 3C**).
- Recognize that dissolution requires: (1) disrupting the intermolecular forces (weak bonds) that hold the network of solvent molecules together, (2) disrupting the intermolecular forces (for covalent compounds) or electrostatic forces (for ionic compounds) that hold the network of solute molecules or ions together, and (3) forming new interactions (weak bonds) between the solute and the solvent (**Exploration 3C**).
- Describe a solvated ion in solution (**Exploration 3C**).

Structure-Solubility Relationships

You should be able to:

- Show why water is a polar molecule (**Exploration 3B**).
- Determine whether or not a molecule is polar, based on the electronegativity of the component atoms and the shape of the molecule (**Explorations 3B and 3C**).
- Recognize that polar solvents dissolve polar substances and non-polar solvents dissolve non-polar substances (**Exploration 3C**).
- Predict the solubility of substances based on their structure and composition (**Explorations 3B and 3C**).

Thermodynamics of Dissolution Reactions

You should be able to:

- Describe dissolution in terms of enthalpy changes (bond making and bond breaking, recognizing that bond breaking requires energy and bond making releases energy) (**Exploration 3C**).

You may wish to look up these topics in your introductory chemistry textbook for additional explanations and examples. They will most likely be discussed in chapters on solution chemistry and thermodynamics and on intermolecular forces.

Thinking Skills

What general skills are you building for your resume?

As you worked through Session 3, you have also been developing some general problem solving and scientific thinking skills that are valued by employers in a wide range of professions and in academia. Here is a list of the skills that you have been building for your resume.

Analogical and Deductive Reasoning Skills

You should be able to:

- Gather and organize data to look for trends and differences (**Exploration 3C**).
- Apply a set of principles learned in one setting to make predictions in a different setting (**Explorations 3B** and **3C**).
- Use existing data to draw analogies to a new situation (**Exploration 3C**).

Checking Your Progress

What progress have you made toward answering the Module Question?

Equilibrium

How can we best describe the extent of a chemical reaction?

Exploration 4A: The Storyline

Ionic substances enter the water supply primarily through the dissolution of minerals. In Session 3, you saw that the solubilities of ionic substances vary. What does this mean for any given substance, especially the ionic compounds in our water supply? Will they all dissolve completely, given enough time? Or will the *extent* to which this chemical reaction proceeds vary as well?

In fact, there are many gradations of reaction "completeness." This may come as a surprise, since we often assume all reactions proceed completely from reactants to products. After all, they are written that way in the chemical equation. Consider the chemical reaction that describes the dissolution of sodium chloride:

$$NaCl\ (s)\ +\ H_2O\ (l) \longrightarrow Na^+\ (aq)\ +\ Cl^-\ (aq)$$

Chemical reactions can also proceed in the reverse direction from the way they are written, so our equation for the dissolution of NaCl can also be written as the precipitation of NaCl:

$$Na^+\ (aq)\ +\ Cl^-\ (aq) \longrightarrow NaCl\ (s)\ +\ H_2O\ (l)$$

We can combine these two reactions as follows:

$$NaCl\ (s)\ +\ H_2O\ (l) \rightleftharpoons Na^+\ (aq)\ +\ Cl^-\ (aq)$$

where the two-way arrows indicate that the forward and reverse reactions are both proceeding to some extent. In this Session, we will look at the processes that are taking place at the atomic and molecular level as reactions proceed in the forward and reverse directions. We will also examine some techniques for quantifying the extent of a reaction.

The goal of this Session is to answer the question, *How can we best describe the extent of a chemical reaction?* We will examine several different reactions, taking note of how far the reaction proceeds as a function of time.

Consider the addition of 50.00 grams of solid sodium chloride to a beaker containing 100 mL of pure water. As time passes, a point is reached at which no more salt dissolves and the solution is **saturated**. For NaCl, the saturation point is reached when 35 grams of NaCl has dissolved in 100 mL of water at room temperature. Table 4-1 shows the masses of solid salt and of dissolved salt as a function of time.

Table 4-1   Concentration Data for Dissolution of NaCl as a Function of Time

Time (s)	Mass of undissolved NaCl (g)	Mass of dissolved NaCl (g)
0	50.00	0
10	40.5	9.5
20	32.6	17.4
30	25.5	24.5
40	20.8	29.2
50	18.6	31.4
60	17.5	32.5
100	15.8	34.2
140	15.2	34.8
200	15.0	35.0
300	15.0	35.0

1. Plot the mass of dissolved and undissolved NaCl (*y* values) as a function of time (*x* value).

2. At what time during the dissolution process has the system reached an equilibrium position? What criteria did you use to determine this?

3. What is true about the ratio of dissolved to undissolved NaCl at equilibrium?

Working with the Ideas

4. How would the plot differ for a *more* soluble substance such as sodium hydroxide, which is soluble to the extent of 42 g/100 mL? Sketch this plot. Assume the same initial conditions of 50 g of solid in 100 mL of water and approximately the same time for the system to reach equilibrium.

5. How would the plot differ for a *less* soluble substance such as calcium carbonate, $CaCO_3$, which is soluble to the extent of only 9.3×10^{-4} g/100 mL? Sketch this plot. Assume the same initial conditions of 50 g of solid in 100 mL of water and approximately the same time for the system to reach equilibrium.

6. In this heterogeneous system (solid/liquid), what factors affect the rate at which the system reaches equilibrium?

Looking Ahead

Several Exploration Questions can guide you to explore the Session Question in more depth. These will be used at the discretion of your instructor.

- **Exploration 4B:** What is equilibrium?

- **Exploration 4C:** What does a chemical system at equilibrium look like at the microscopic level?

- **Exploration 4D:** How can equilibrium reactions be described mathematically?

- **Exploration 4E:** How can we use the equilibrium expression to predict equilibrium concentrations?

- **Exploration 4F:** How can you tell whether a reaction has reached equilibrium?

- **Exploration 4G:** How is free energy related to the extent of a reaction?

Exploration 4B

What is equilibrium?

Creating the Context

Why is this an important question?

_____

What do you already know?

_____

In Exploration 4A, you saw that sodium chloride dissolves in water, but only to a certain extent. Once the maximum amount of salt has dissolved, the system reaches **equilibrium**, the point at which no obvious changes in concentration occur with time. All chemical reactions can be described as equilibria under certain conditions, and by knowing more about equilibrium, we can quantify and predict the extent of any chemical reaction.

The idea of equilibrium is probably familiar to you. As a child, you may have played on a see-saw with a friend of approximately equal weight. If you both stopped pushing off the ground, the see-saw would move up and down and slowly come to rest. At this point it was exactly balanced if you weighed the same as your friend, or a bit tilted if one person was slightly heavier than the other, or sitting closer to the end. In either case, your see-saw had reached an equilibrium position.

There are two types of equilibrium, **static** and **dynamic**. The see-saw example is a **static** equilibrium: all forces are balanced and no changes occur once the system has reached equilibrium. In a **dynamic** equilibrium, change is occurring continuously in opposing directions, but the net result of the opposing changes is zero. To an outside viewer, a system in dynamic equilibrium does not appear to change.

Developing Ideas

1. Begin by writing down your own definition of equilibrium.
2. Classify the following equilibrium situations as *static* or *dynamic*.
    *   A rock lying on the ground.
    *   A kayaker staying in place on a flowing river by paddling upstream.
    *   A tug-of-war between two equally matched teams.
    *   A person walking on a treadmill.
3. Go to the CD or web-site to find the Equilibrium Simulator (http://chemistry.beloit.edu/Water/equilibK/index.html). Observe the computer animation of a chemical reaction at equilibrium, where the reactants are represented by small red spheres, the products by larger blue spheres.

$$2\,A \rightleftharpoons B$$
$$\text{red} \qquad \text{blue}$$

Would you classify a chemical equilibrium as static or dynamic? Explain.

Working with the Ideas

The following problems will help you understand the concepts in more depth.

4. For each of the equilibria you classified as dynamic in problem 2:
    a. Identify the two opposing processes.
    b. How does the rate of change in one direction compare to the rate of change in the opposite direction?
5. In the example of salt (NaCl) dissolving in water, what are the two opposing processes that make up the dynamic equilibrium?

Exploration 4C

What does a chemical system at equilibrium look like at the microscopic level?

Creating the Context

Why is this an important question?

Chemical equilibria play an important role in removing most contaminants from water supplies. Thus, we will need to understand more about equilibrium to make the best decisions about cleaning up a water supply. Before we go much further, we need to examine the conditions under which a chemical system can reach equilibrium.

a. The system containing reactants and products must be closed, such that the system does not gain or lose any product or reactant molecules.

b. The temperature must be constant.

c. The pressure in the system must be constant.

d. There should be no other chemical reactions or physical processes that will use up or generate the products or reactants.

We will explore non-equilibrium situations in more depth in Session 5. In this Session, we will examine the dynamic process of chemical equilibrium and learn how the *rates* of the forward and reverse reactions change as a system approaches equilibrium. Knowing that chemical equilibria are dynamic will help us build a working model of the reactions at the atomic and molecular level. Our goal will be to answer the question *What does a chemical system at equilibrium look like at the microscopic level?*

Preparing for Inquiry

BACKGROUND READING

The end result of an equilibrium chemical reaction is a mixture containing both reactants and products, with the relative amounts of reactants and products dependent on the particular reaction.

$$aA \ + \ bB \ \rightleftharpoons \ cC \ + \ dD$$

On the microscopic level, collisions between atoms are required for a reaction to proceed in either the forward or reverse direction. As atoms collide, bonds are broken and formed as reactants are converted to products. After some products have accumulated, they suffer collisions as well and are converted back to reactants. The number of collisions that occur in a particular system depends on several factors:

- *The number of reactant and product atoms or molecules present per unit volume, also known as the concentration.* Higher concentrations of reacting particles lead to more collisions and faster reaction rates. As an analogy, think of a dining hall where people move from the food line to the drink machine to the station with the utensils to the cash register and finally to their seats. If just a few people are present, the likelihood of colliding with one of them would be small. However, if the dining hall is crowded, the probability of colliding with another person is much larger.

- *The temperature.* As temperature increases, atoms and molecules move faster, because they possess greater kinetic energy. As a result, the number of collisions in a given time period increases, and thus the rate of the reaction increases. Using the dining hall analogy again, consider what would happen if everyone were *running* from place to place instead of walking. Certainly more collisions would occur.

With these ideas in mind, we need to look at the forward and reverse reactions more carefully. The rates of these reactions depend on the temperature, as well as on the concentrations of reactants and products. For the generic reaction that follows,

$$aA \; + \; bB \; \underset{k_r}{\overset{k_f}{\rightleftharpoons}} \; cC \; + \; dD$$

the rates of the forward and reverse reactions can be expressed as a function of concentration at a particular temperature by the following equation:

$$\text{Rate (forward rxn.)} = k_f \, [A]^a [B]^b$$

$$\text{Rate (reverse rxn.)} = k_r \, [C]^c \, [D]^d$$

A constant is associated with each reaction (k_f for the forward reaction and k_r for the reverse reaction); its value is related to the likelihood that a transformation will occur when particles collide. Rates can also be expressed as a derivative, or a change in concentration of one of the species with respect to a change in time (e.g., $-d[A]/dt$, a decrease in the concentration of A over time, or $d[C]/dt$, an increase in the concentration of C over time).

How can you find out more?

One way to learn more about these dynamic processes and how they relate to equilibrium is to observe some chemical reactions and carefully evaluate the change in concentrations of reactants and products over time. The **Equilibrium Simulator** is a computer model of a chemical reaction that allows you to take a microscopic look at reaction dynamics.

HOW TO USE THE EQUILIBRIUM SIMULATOR

To use the **Equilibrium Simulator**, use the CD-ROM or go to the web site http://chemistry.beloit.edu/Water/equilibK/index.html.

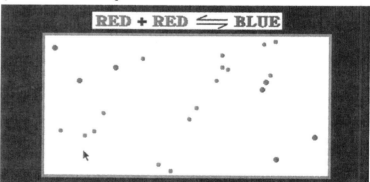

The Equilibrium Simulator allows you to experiment with the variables controlling reaction rates and the extent of a chemical reaction. You can change the following values:

- **Red Number:** Allows you to change the number of red particles, from a minimum of zero to a maximum of 40.

- **Blue Number:** Allows you to change the number of blue particles, from a minimum of zero to a maximum of 40.

NOTE: Once the simulator begins to run, the numbers displayed in the Red Number and Blue Number boxes represent a running average of the numbers of red and blue particles in the system.

- **Temperature:** Allows you to change the temperature and thus the speed of the particles. Higher temperatures mean faster speeds and more collisions between particles.

- **Probability:** The likelihood that a reaction will occur when two particles collide. *Do not change this parameter for these exercises!*

To operate the Simulator, enter the desired values for temperature and numbers of particles and then click on **Equilibrate!** to start the "reaction." Click on **Pause** to stop the reaction. For best performance, turn off all other software applications.

Developing Ideas

We will begin by considering the chemical reaction discussed in Exploration 4B, that of two molecules of A reacting to form B.

$$2\,A \;\rightleftharpoons\; B$$
$$\text{red} \qquad \text{blue}$$

1. Enter the following values and click on **Equilibrate!** to start the reaction.

Red Number	5
Blue Number	0
Temperature	273

   a. At the instant the reaction begins, what is the rate of the reverse reaction, that of B going to A? Explain in the context of a microscopic view of the reaction and collisions of molecules.

   b. List two ways you might be able to increase the reaction rate. Try them out and see if you are correct.

2. Enter the following values and click on **Equilibrate!** to start the reaction. Allow

	Start	Equilibrium
Red Number	14	
Blue Number	14	
Temperature	273	

the reaction to proceed, observing the change in the number of red and blue particles. Remember that the Red Number and Blue Number displays a running average of the number of red and blue particles, so the number in the box will not necessarily represent the exact number of red or blue particles at a particular instant.

   a. How does the number of red particles change as a function of time over the first 30 seconds? Blue particles?

   b. How does the number of red particles change as a function of time over the next 30 seconds? Blue particles?

   c. How does the number of red particles change as a function of time over the next minute? Blue particles?

   d. At approximately what time would you say that the reaction has reached equilibrium? Explain how you know. Enter the equilibrium values for Red Number and Blue Number in the table above.

3. For this exercise, you will examine two other sets of values.

Set 1	Start	Equilibrium	Set 2	Start	Equilibrium
Red Number	0		Red Number	40	
Blue Number	21		Blue Number	0	
Temperature	273		Temperature	273	

For each set of values, click on **Equilibrate!** and allow the reaction to proceed until no further change occurs in the numbers of red and blue particles.

a. Write down the equilibrium value for Red Number and Blue Number in the table above.

b. Compare these values to those you obtained in Problem 2 above. How is the position of the equilibrium affected by the starting number of particles?

Working with the Ideas

4. If you were to start with only red particles (call them **A**) and plot the rate of change in concentration of **A** and **B** (blue particles) vs. time as the system approaches equilibrium, you would obtain a plot similar to that in Figure 4-1. Use the diagram and the Equilibrium Simulator to determine whether each of the following statements is true or false. Explain your reasoning.

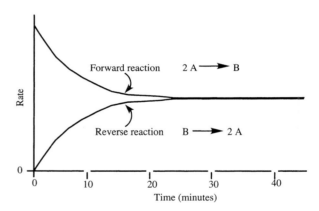

Figure 4-1:   A plot of the rates of the forward and reverse reactions vs. time for a reaction starting with only **A**.

Initially, at time = 0 minutes:

a. The rate of the forward reaction exceeds the rate of the reverse reaction.

b. The rate of the reverse reaction exceeds the rate of the forward reaction.

c. For a period of time after initial mixing, the concentration of the products increases.

d. For a period of time after initial mixing, the concentration of the reactants increases.

At equilibrium:

a. The rate of the forward reaction is zero.

b. The rate of the reverse reaction is zero.

c. The rate of the forward reaction is equal to the rate of the reverse reaction.

d. The rates of the forward and reverse reactions are both constant.

5. Which of the following systems are at equilibrium? Explain your reasoning.

    a. Minerals dissolving into flowing stream water

    b. Gasoline burning in a car engine

    c. A capped bottle of brine solution, with some solid salt remaining in the bottom of the bottle

    d. A closed soft drink at room temperature that has just been placed in a cooler

---

Exploration 4D

How can equilibrium reactions be described mathematically?

Creating the Context

Why is this an important question?

What do you already know?

How can you find out more?

Some water treatment processes involve adding substances to source water to initiate chemical reactions with impurities. These reactions may remove the impurities or render them harmless. However, you now know that chemical reactions do not all proceed to the same extent. How do water treatment engineers know how much of a chemical to add to ensure sufficient removal of a contaminant?

In Exploration 4C, you observed an animation of a chemical reaction at equilibrium. What you should have noticed is that, even though the red and blue particles representing the reactants and products interconvert, the total number of red particles (reactant) and the total number of blue particles (product) remained approximately constant over time. In a system containing a greater number of particles, it is not the absolute number that remains constant but the *ratio* of products to reactants. So we'll need to use mathematics to answer the question about how far a given reaction proceeds.

The goal of this Session is to answer the question *How can equilibrium reactions be described mathematically?* This Exploration will provide you with the tools to do this math and obtain answers to equilibrium problems.

Developing Ideas

A chemical system at equilibrium has a constant concentration of products and reactants; however, the relationship between products and reactants is a bit more complex than a simple ratio of concentrations. Your goal in this exercise is to determine more precisely the relationship between the concentrations of products and reactants in a reaction at equilibrium and to determine the general mathematical form of the **equilibrium expression** that yields a constant value at a given temperature.

1. Consider a simple chemical reaction, an equilibrium between nitrogen dioxide (NO_2) and dinitrogen tetroxide (N_2O_4). Both of these compounds are gases.

    a. Write a balanced chemical equation for the equilibrium between NO_2 and N_2O_4. While you can write the reaction in either direction, for the purposes of this exercise, assume that NO_2 is the *product* and N_2O_4 is the *reactant*.

    b. Using the data in Table 4-2, fill in the empty columns by calculating the ratios of concentrations shown in the table headings.

    c. Which of the calculated ratios of concentrations can best be described as a constant?

    d. Compare the mathematical form of the constant ratio to the balanced chemical equation you wrote in part *a*. What relationship exists between the stoichiometric coefficients of the balanced equation and the power to which the concentration is raised in the ratio?

Table 4-2 Concentration Data for the NO_2-N_2O_4 System at 25°C

Initial concentrations* (M)		Equilibrium concentrations* (M)		Ratios of equilibrium concentrations*		
$[NO_2]$	$[N_2O_4]$	$[NO_2]$	$[N_2O_4]$	$\dfrac{[NO_2]}{[N_2O_4]}$	$\dfrac{[NO_2]^2}{[N_2O_4]}$	$\dfrac{[NO_2]+[NO_2]}{[N_2O_4]}$
0.00	0.67	0.055	0.64	0.085		0.17
0.050	0.45	0.046	0.45		0.0047	
0.030	0.50	0.048	0.49			
0.040	0.60	0.052	0.59	0.088		0.18
0.20	0.00	0.021	0.091		0.0048	

*Note: The square brackets indicate concentrations in moles per liter.

    2. Apply what you learned in question 1 to write a general formula for expressing the ratios of concentrations at equilibrium for the generic chemical reaction that follows. This ratio is called the **equilibrium expression**.

$$aA \ + \ bB \ \rightleftharpoons \ cC \ + \ dD$$

    3. Think about the numerical value of the constant you obtained for the reaction of N_2O_4 to form NO_2. Do you think this reaction proceeds to a large extent or to a small extent? Explain.

Working with the Ideas

The following problems will help you understand the concepts in more depth. As you practice writing equilibrium expressions, keep the following rules in mind.

- Concentrations can be expressed in one of several ways:
  Gases: Use moles of solute per liter of solution (equilibrium constant is called K_C) or pressure in atmospheres (equilibrium constant is called K_P).
  Components in solution: Use moles per liter
  Pure liquids or solids: Do not enter into the equilibrium expression because their concentrations are essentially constant.

- Equilibrium constants are usually expressed as unitless numbers. See Appendix 4A for more details.

    4. Write the equilibrium expressions for the following equilibria using concentrations in moles per liter.

      a.

$$2\, CO_2\,(g) \rightleftharpoons 2\, CO\,(g) \ + \ O_2\,(g)$$

b.

$$HCl\,(g)\ +\ H_2O\,(l)\ \rightleftharpoons\ H_3O^+\,(aq)\ +\ Cl^-\,(aq)$$

c.

$$2\,NO_2\,(g)\ +\ 7\,H_2\,(g)\ \rightleftharpoons\ 2\,NH_3\,(g)\ +\ 4\,H_2O\,(l)$$

5. Hydrogen gas (H_2) reacts with iodine gas (I_2) to form an equilibrium mixture of H_2, I_2, and HI.

    a. Write a balanced chemical reaction that describes this equilibrium.

    b. Write the equilibrium expression for the reaction.

    c. The value of the equilibrium constant for the reaction with HI as the product is 54.3 at 430°C. At equilibrium, will there be more HI or more H_2 and I_2? Explain how you know.

    d. How would the equilibrium expression for the *decomposition* of HI into H_2 and I_2 be different from the one you wrote in b? How will the value of the equilibrium constant change?

6. Consider the reaction of N_2O_4 to form two molecules of NO_2 at 25°C. The equilibrium constant for the reaction is 0.0046. Sketch a rough plot of concentration vs. time for the following situations.

    a. Starting with only NO_2

    b. Starting with only N_2O_4

    c. Starting with a mixture of NO_2 and N_2O_4

7. The following reaction reaches equilibrium:

$$NH_3\,(aq)\ +\ H_2O\,(l)\ \rightleftharpoons\ NH_4^+\,(aq)\ +\ OH^-\,(aq)$$

    a. Write the equilibrium expression for the reaction.

    b. What parameters would you have to measure to determine the value of the equilibrium constant experimentally?

    c. The equilibrium constant for this reaction is less than 1. What does this tell you about the relative concentrations of the products and reactants?

    d. If the equilibrium constant for a reaction were equal to 1, what can be said mathematically about the relative concentrations of the products and reactants?

8. The equilibrium expression contains very similar quantities to the rate expressions shown on page 76. Assuming that when a system is at equilibrium, the rate of the forward reaction equals the rate of the reverse reaction, manipulate the rate equations algebraically to give the equilibrium expression.

Exploration 4E

How can we use the equilibrium expression to predict equilibrium concentrations?

In water chemistry, many equilibria are occurring at any given time. Minerals are dissolving, water treatment chemicals are reacting with impurities, and atmospheric gases are dissolving—all to a greater or lesser extent, depending on the reaction. To purify our water, we'll need to predict the extent to which these reactions occur.

As you discovered in Exploration 4D, the equilibrium expression and accompanying equilibrium constant provide a mathematical way to determine the ratio of the concentrations of products to the concentrations of reactants. For example, a lake that contains calcite rock will be rich in calcium because calcite dissolves by the following reaction:

$$CaCO_3 \text{ (s)} \xrightleftharpoons{K_{eq}} Ca^{2+} \text{ (aq)} + CO_3^{2-} \text{ (aq)}$$

We can use our knowledge of equilibrium and the constant ratio of products to reactants to predict the concentration of dissolved species in the lake. When the system reaches equilibrium, it is saturated, meaning that the maximum amount of solid has dissolved at that temperature.

When a system is at equilibrium, it can be mathematically described by an equilibrium expression that is a ratio of the concentrations of products to reactants. For the preceding reaction at 25°C:

$$K_{eq} = \frac{\left[CO_3^{2-}\right]\left[Ca^{2+}\right]}{\left[CaCO_3(s)\right]}$$

This expression can be simplified by observing that the concentration of the solid is a constant and does not vary with the amount of solid present. In fact, the amount of solid present is completely unrelated to how much of the compound dissolves, *as long as some solid is present*. As you may recall from **Session 3**, the concentration of a dissolved ionic compound at equilibrium is related to how strong the ion-ion bonds are relative to the ion-water bonds formed when the compound dissolves. Thus as long as some solid is present, the compound will dissolve to the extent governed by the energetics of the process.

One useful outcome of this fact is that the equilibrium expression can be simplified by eliminating the undissolved solid from the equilibrium expression for a particular reaction. Thus, for the dissolution reaction of calcium carbonate, the entire equilibrium expression is simplified to give a new expression that we designate as the **solubility product constant, K_{sp}**, where

$$K_{sp} = \left[Ca^{+2}\right]\left[CO_3^{-2}\right] = 8.3 \times 10^{-9}$$

Tables of K_{sp} values are readily available (see Appendix 4B).

Chemists use two different terms to describe how much of a compound will dissolve in a given volume of water:

- The **molar solubility (s),** or the number of *moles* of a compound that will dissolve in a liter of water to form a saturated solution. Recall that molarity is the concentration unit used for components in solution.

- The **solubility**, or the number of *grams* of a compound that will dissolve in a liter of water to form a saturated solution.

In this Exploration, we will examine how we can use this information and our knowledge of reaction stoichiometry to predict equilibrium concentrations in water supplies and in water treatment reactors. Thus, the goal of this Exploration is to answer the question *How can we use the equilibrium expression to predict equilibrium concentrations?*

Developing Ideas

Slaked lime, $Ca(OH)_2$, is a slightly soluble salt used in water treatment reactions as a source of hydroxide ions.

$$Ca(OH)_2\ (s) \overset{K_{sp}}{\rightleftharpoons} Ca^{2+}\ (aq)\ +\ 2\ OH^-\ (aq)$$

Hydroxide ions react with metals, such as iron, to precipitate out metal-hydroxide salts. Let's evaluate the dissolution of lime, $Ca(OH)_2(s)$, dissociating to form aqueous calcium ions (Ca^{2+}) and hydroxide ions (OH^-). We will begin by mixing $Ca(OH)_2(s)$ and water and call that our *Initial* condition. As time passes, part of the $Ca(OH)_2$ dissolves, resulting in a solution of calcium and hydroxide ions. When the system reaches equilibrium, we will have a solution of calcium and hydroxide ions with solid calcium hydroxide remaining on the bottom of the flask (see Figure 4-2). We will call this the *Final* condition.

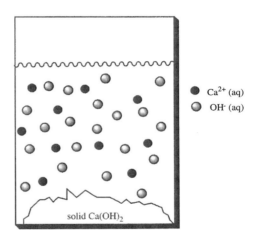

Figure 4-2: A representation of the dissolution of $Ca(OH)_2$ in water. For simplicity, water molecules are not shown and the sizes of the dissolved ions relative to the size of the container are greatly exaggerated.

1. What is the stoichiometric relationship between the number of moles of $Ca(OH)_2(s)$ dissolved and the number of moles of Ca^{2+} and OH^- ions formed?

2. We don't know yet exactly how many Ca^{2+} and OH^- ions actually dissolved in the solution, but we can represent this unknown amount with an x. Because you know the relationship between the moles of $Ca(OH)_2$ dissolved and the moles of Ca^{2+} and OH^- ions produced, you now have a system with just one unknown.

Fill in the following table, using the reaction stoichiometry and x for any unknown amount of aqueous Ca^{2+} and OH^- ions.

	$Ca(OH)_2(s)$ (mol/L)	Ca^{2+} (aq) (mol/L)	OH^- (aq) (mol/L)
Initial Conc.	constant	0	0
Final Equilibrium Concentration in terms of x (mol/L)	constant		

3. The solubility product constant, K_{sp}, for this reaction is 5.5 x 10^{-6} at 25°C.
    a. Write the equilibrium expression for the dissolution reaction.
    b. Substitute the final equilibrium concentrations from the table above into the equilibrium expression.
    c. Solve for x to obtain the concentrations of both Ca^{2+} and OH^- at equilibrium in moles per liter.
    d. Calculate both the *molar solubility* and the *solubility* of $Ca(OH)_2$.

Working with the Ideas

The following problems will give you more practice with the concepts.

4. The equilibrium expression for the dissolution of $Fe(OH)_3(s)$ to form aqueous Fe^{3+} and OH^- ions takes the following form:

$$K_{sp} = 4 \text{ x } 10^{-38} = 27x^4$$

Show how this expression was obtained based on the stoichiometry of the dissolution reaction.

5. You are trying to obtain a certain concentration of hydroxide ion in a solution. The concentration produced in the equilibrium dissociation of solid $Ca(OH)_2$ is 0.011 M. If you wish to have a higher concentration of hydroxide in the solution, which of the following methods would work? Explain your reasoning.
    a. Add more solid $Ca(OH)_2$
    b. Add sodium hydroxide, NaOH ($K_{sp} \gg 1$)
    c. Add magnesium hydroxide, $Mg(OH)_2$ ($K_{sp} = 1.2$ x 10^{-11})

6. The relative molar solubility of two substances can *sometimes* be predicted by evaluating K_{sp} values and thinking about the relative amounts of products and reactants and the position of the equilibrium.

a. Look at the following table and, without doing a calculation, predict which compound of each pair would have the higher molar solubility.

Compound	K_{sp}	Solubility Prediction (soluble/insoluble)	Calculated Result
$CaSO_4$	1.9×10^{-4}		
$CaCO_3$	8.7×10^{-9}		
CaF_2	4.0×10^{-11}		
$Ca(OH)_2$	5.5×10^{-6}		
$CaCO_3$	8.7×10^{-9}		
CaF_2	4.0×10^{-11}		
$CaCO_3$	8.7×10^{-9}		
$Ca_3(PO_4)_2$	2.0×10^{-29}		

b. Now go back and do a calculation to determine the exact molar solubility of each compound. Where do your predictions NOT match the actual solubilities? (*Hint:* Write out the equation for the solubility reaction and take a close look at the stoichiometric coefficients in the equation.)

c. Generalize your result. Under what circumstances can you use the values of K_{sp} to compare molar solubilities?

7. The solubility of $CaSO_4$ is found to be 0.67 g/L. Use this information to find the K_{sp} for $CaSO_4$. Show your work.

Exploration 4F

How can you tell whether a reaction has reached equilibrium?

Creating the Context

Why is this an important question?

In Exploration 4E, we used equilibrium expressions to calculate the amount of hydroxide present when a certain amound of lime was added to a water supply. But this process also *adds* calcium ions, one of the contaminants that we want to *remove*, because it adds to water hardness. So we now have to deal with natural sources of calcium as well as the calcium we've added for other reasons. One of the main reactions used to remove calcium from water is precipitation of the calcium as calcium carbonate ($CaCO_3$).

$$Ca^{2+} (aq) \ + \ CO_3^{2-} (aq) \ \rightleftharpoons \ CaCO_3 (s)$$

Soda ash (Na_2CO_3) is usually used as the source of carbonate ions. This precipitation reaction can take hours, so to ensure adequate removal of the calcium, you must have some way to determine when the reaction is nearly complete.

What background information is useful?

To accomplish this goal, we must use the equilibrium expression in a slightly different way and define a new ratio of products to reactants, the **reaction quotient**, Q. The reaction quotient expression takes the same form as the equilibrium expression, but instead of describing concentrations at *equilibrium*, it describes concentrations at *any moment in time*. For the dissolution of calcium carbonate (the *reverse* of the reaction shown above:

$$Q = \left[Ca^{2+}\right]\left[CO_3^{2-}\right]$$

In this Exploration, the goal is to learn to use the reaction quotient Q to answer the question *How can you tell whether a reaction has reached equilibrium?*

Developing Ideas

You are a water treatment engineer who is removing calcium from a water supply with an initial calcium concentration of 0.0075 M. A stoichiometric amount of sodium carbonate is added to precipitate the calcium as solid $CaCO_3$ at 25°C.

$$Ca^{2+}\ (aq)\ +\ CO_3^{2-}\ (aq)\ \overset{K'}{\rightleftharpoons}\ CaCO_3\ (s)$$

1. Write the equilibrium expression for the *precipitation* of $CaCO_3$.
2. Knowing that the K_{sp} for the dissolution of $CaCO_3$ is 8.7×10^{-9}, determine the equilibrium constant for the *precipitation* of $CaCO_3$. [*Hint:* Compare the two equilibrium expressions for the reaction written as a dissolution (K_{sp}) and the reaction written as a precipitation (K').]
3. What will be the equilibrium concentration of Ca^{2+} when the maximum amount of $CaCO_3$ has dissolved in water? What is the value of Q for the dissolution of $CaCO_3$ at this point?
4. Use the following data to calculate the missing reaction quotients for the precipitation reaction at different times. When has the system reached equilibrium?

Time (hours)	[Na$^+$] (mol/L)	[CO$_3^{2-}$] (mol/L)	[Ca^{2+}] (mol/L)	Reaction Quotient (Q)
0	1.7×10^{-2}	8.5×10^{-3}	7.5×10^{-3}	1.6×10^4
0.25	1.7×10^{-2}	3.5×10^{-3}	2.5×10^{-3}	
0.5	1.7×10^{-2}	1.8×10^{-3}	8.0×10^{-4}	
1	1.7×10^{-2}	1.1×10^{-3}	2.3×10^{-4}	4.0×10^6
4	1.7×10^{-2}	1.1×10^{-3}	5.4×10^{-5}	
8	1.7×10^{-2}	1.1×10^{-3}	1.2×10^{-5}	
12	1.7×10^{-2}	1.1×10^{-3}	7.9×10^{-6}	1.2×10^8
16	1.7×10^{-2}	1.1×10^{-3}	7.9×10^{-6}	

5. Why does the sodium concentration remain the same throughout the reaction?
6. How does the final concentration of Ca^{2+} compare to the concentration of Ca^{2+} calculated in problem 3? Explain why these are not the same.
7. What generalizations can you make about Q and K_{eq}? Choose the correct response from the choices given.
    a. When $Q > K_{eq}$, the equilibrium position of the reaction
        ____shifts to the left ____shifts to the right ____remains the same
    b. When $Q < K_{eq}$, the equilibrium position of the reaction
        ____shifts to the left ____shifts to the right ____remains the same

c. When $Q = K_{eq}$, the equilibrium position of the reaction
    ___shifts to the left ___shifts to the right ___remains the same

8. You treat 100,000 liters of water having $[Ca^{2+}] = 0.0075$ M with 10 grams of Na_2CO_3 but see no precipitate of $CaCO_3$. Why not?

Working with the Ideas

The following problems will give you more practice with the concepts.

9. When it rains, water flows over rocks and soil, slowly dissolving the slightly soluble minerals. Water that collects in lakes and streams has a characteristic chemical signature based on the minerals in the nearby rocks and soils. Two common minerals are calcite ($CaCO_3$, $K_{sp} = 2.8 \times 10^{-9}$) and gypsum ($CaSO_4$, $K_{sp} = 9.1 \times 10^{-6}$).

a. In an area containing predominantly gypsum rock, what equilibrium concentrations of Ca^{2+} and SO_4^{2-} would you expect to find in the water in moles per liter? In mg/L (ppm)?

b. When the water was tested after a heavy rain, the *actual* concentration of Ca^{2+} in the water was found to be 0.0012 M. Compare this concentration with the one you obtained for Question 9a and explain why the numbers might be different.

10. Water pipes made of lead will react with carbonate (CO_3^{2-}) in the water to form lead carbonate ($PbCO_3$) on the inside of the pipes. When a water faucet is unused for long periods of time, the $PbCO_3$ in contact with the water in the pipe dissolves to the extent governed by its equilibrium constant, forming toxic Pb^{2+} ions and CO_3^{2-} ions. What is the concentration of Pb^{2+} in drinking water that comes out of such a faucet? The K_{sp} for $PbCO_3$ is 3.3×10^{-14}.

11. Public information pamphlets advise people who suspect that lead is in their water to have it tested and, if necessary, replace the pipes. In the meantime, they advise you to run the water for 15-30 seconds before using it, and to use only cold water. Why might this be? Why is it not a permanent solution?

12. You find that the reaction of Ca^{2+} with CO_3^{2-} to precipitate out $CaCO_3$ goes faster if you add an excess of sodium carbonate. You decide to try several different concentrations of sodium carbonate to see which one works best. Use the data in the following table to determine which of the reactors have reached equilibrium. Explain your reasoning and show your work.

Reactor	$[Ca^{2+}]$ (mol/L)	$[CO_3^{2-}]$ (mol/L)	Reaction Quotient (Q)
1	7.5×10^{-3}	1.2×10^{-2}	
2	2.5×10^{-5}	3.5×10^{-3}	
3	7.0×10^{-8}	1.2×10^{-1}	

13. A lake contains magnesium carbonate ($MgCO_3$) bedrock. What are the equilibrium concentrations of Mg^{2+} and CO_3^{2-} in the lake?

$$MgCO_3 \text{ (s)} \rightleftharpoons Mg^{2+} \text{ (aq)} + CO_3^{2-} \text{ (aq)} \qquad K_{sp} = 4.0 \times 10^{-5}$$

14. A water supply contains 1 mg/L of iron as Fe^{2+}, which is more than three times higher than the MCL. You choose to remove the iron by precipitation with hydroxide ions, OH^-.

$$Fe^{2+} (aq) \ + \ 2 OH^- (aq) \ \rightleftharpoons \ Fe(OH)_2 (s)$$

a. What is the minimum amount of solid NaOH (in grams) you would have to add to 100,000 liters of water to observe precipitation of $Fe(OH)_2$?

b. What is the stoichiometric amount of NaOH required to react with all of the iron in the 100,000 liters?

c. If you add a stoichiometric amount of NaOH, what is the final concentration of iron in the water in moles per liter? In mg per liter? Does it meet the MCL for iron? What is the final concentration of Na^+ in mg/liter? Does it exceed the MCL for Na^+?

Exploration 4G

How is free energy related to the extent of a reaction?

Creating the Context

Why is this an important question?

If you have already studied thermodynamics, you may have noticed the similarities between the words we use to describe the free energy changes of a reaction and the extent of a reaction. "Product-favored" and "reactant-favored" are used to describe both the equilibrium state and the energetics of a chemical reaction. This terminology suggests that these two key descriptors of chemical reactions, ΔG and K (or Q), are related. Indeed, there is such a relationship and it can be described by the following equation:

$$\Delta G = \Delta G^\circ + RT \ln Q$$

Although we will not examine the origin of this equation in this module, we will look at what the equation tells us about the relationship of free energy to equilibrium and the extent of a reaction. We can take advantage of the extensive database of thermodynamic data to determine equilibrium constants without having to measure them in the lab. Conversely, we can also use equilibrium constants to determine reaction free energies.

What do you already know?

From your background in thermodynamics, remember that the standard free energy change, ΔG°, is the free energy change for a reaction taking place under standard conditions, where all concentrations are 1 M (for components in solution) or 1 atmosphere (for gases) and at a constant temperature, 25°C in most tabulations. The standard free energy change for a reaction (ΔG°_{rxn}) can be calculated from the standard free energy of formation (ΔG°_f) of the reactants and products, as follows:

$$\Delta G^\circ_{rxn} = \Sigma\, \Delta G^\circ_f(\text{products}) - \Sigma\, \Delta G^\circ_f(\text{reactants})$$

In contrast, ΔG (with no superscript zero) is the free energy change for a reaction taking place under _any other_ conditions of concentration, pressure, and/or temperature. Since reactions rarely take place under standard conditions, especially if we are adding chemicals to a water supply, we are more often concerned with finding out about ΔG.

In this Exploration, we will examine a variety of reaction scenarios to understand better the relationship between reactant and product concentrations, free energy changes, and the value of the equilibrium constant for a given reaction, with the goal of answering the question _How is free energy related to the extent of a reaction?_

Developing Ideas

Let's begin by looking at a simple reaction, that of reactant, compound A, in equilibrium with product, compound B. The reaction has an equilibrium constant of 3.

$$A \underset{}{\overset{K_{eq}}{\rightleftarrows}} B \qquad\qquad K_{eq} = \frac{[B]}{[A]} = 3$$

1. Consider an initial condition with [A] = [B] = 1 M, i.e., standard state conditions (Figure 4-3).

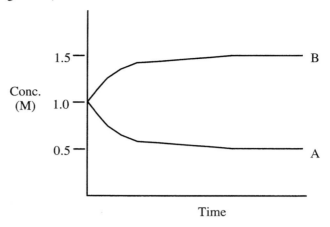

Figure 4-3:   A plot of the concentrations of A and B as a function of time, beginning with all concentrations equal to 1.0 M.

a. What is the value of Q when the reaction begins?

b. Is the reaction at equilibrium under these conditions?

c. Under these conditions, in which direction is the reaction "spontaneous" in the thermodynamic sense, the forward or the reverse direction?

d. What is the sign of ΔG at the beginning of the reaction?

e. When this reaction is at standard state conditions, is it at equilibrium? Explain.

2. An integrated view of the relationship of the reaction energetics and the equilibrium is shown by a plot of ln Q vs. ΔG, which is simply a graphical description of the equation $\Delta G = \Delta G° + RT \ln Q$. Sample data for the reaction described in problem 1 are given in the following table, and Figure 4-4 is a plot of the data. Use the plot and the data table to answer the following questions.

[A] (mol/L)	[B] (mol/L)	$\dfrac{[B]}{[A]} = Q$	ln Q	ΔG (kJ/mol)
1.0	1.0	1.0	0.0	-2.7
0.8	1.2	1.5	0.4	-1.7
0.5	1.5	3.0	1.1	0
0.3	1.7	5.7	1.7	1.6
0.1	1.9	19	2.9	4.6

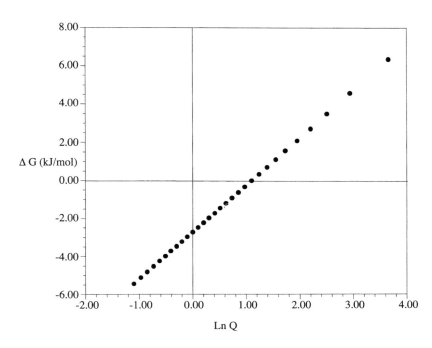

Figure 4-4: A plot of ln Q vs. ΔG shows the relationship between reaction free energy and the reaction quotient.

a.   When the reaction has reached equilibrium, what is the value of ΔG? What does this tell you about the driving force for the reaction to proceed from A to B at this point?

b.   At what point on the graph does $\Delta G = \Delta G°$? Is the reaction at equilibrium at this point?

c.   In what region of the plot is the reaction spontaneous as written (A—>B)?

d.   How would you change the concentrations of A and B to make ΔG, the driving force for the reaction to occur, more negative?

3.   What does the term "product-favored" tell you about

a.   the expected sign of ΔG?

b.   the relative potential energies of products and reactants?

c.   the expected value of the equilibrium constant?

Working with the Ideas

Consider the same reaction, that of reactant, compound A, in equilibrium with product, compound B, under several different initial concentration conditions. The reaction has an equilibrium constant of 3.

$$A \underset{}{\overset{K_{eq}}{\rightleftharpoons}} B \qquad\qquad K_{eq} = \frac{[B]}{[A]} = 3$$

4.   Sketch a plot of concentration vs. time for the following starting conditions:

a.   1 M A and 0.1 M B

b.   0.1 M A and 1 M B

c.   1.5 M A and 0.5 M B

5.   Are any of the preceding reactions at standard state conditions at any time during the approach to equilibrium?

6. Calculate ln Q for the three starting concentrations given in problem 3. Refer to Figure 4-4 to determine the driving force (ΔG) for the reaction to occur as written (A—>B) under these three sets of starting concentrations.

Making the Link

Looking Back: What have you learned?

As you worked through Session 4, you gained experience with the following chemical principles. Understanding these principles can help you solve a wide range of real-world problems. Your experience with Session 4 should have developed your skills to do the items on the following lists.

Chemical Equilibrium

You should be familiar with:

*   The idea that reactions do not necessarily proceed completely from reactants to products. Reactions can proceed to different extents, reaching an equilibrium position (**Exploration 4A**).

*   The idea of chemical equilibrium as a dynamic process, with both the forward and reverse reactions occurring simultaneously. Once at equilibrium, a reaction proceeds in both the forward and reverse directions at identical rates, with a net zero change in the concentrations of products and reactants (**Explorations 4B, 4C**).

Microscopic View

You should be able to:

*   Determine whether or not a system could reach equilibrium under the existing conditions (**Exploration 4C**).

*   Describe a system at equilibrium in terms of the behavior of the atoms, ions, and/or molecules involved in the equilibrium reaction (**Exploration 4C**).

*   Discuss how concentration and temperature affect the rate of a reaction in terms of the behavior of the atoms, ions and/or molecules involved (**Exploration 4C**).

*   For a system approaching equilibrium, show how the rates of the forward and reverse reactions change over time and explain why in terms of the behavior of the atoms, ions, and/or molecules involved (**Exploration 4C**).

Equilibrium Expressions

You should know:

*   How to write equilibrium expressions (**Exploration 4D**). For the reaction

$$aA + bB \rightleftharpoons cC + dD$$

The equilibrium expression is written as

$$K = \frac{[C]^c [D]^d}{[A]^a [B]^b}$$

- The equilibrium expression pertains to a particular chemical equation. The actual value of the equilibrium constant and the form of the equilibrium expression depend on the direction in which the reaction is written and the stoichiometric coefficients of the balanced equation (**Exploration 4D**).

- The value of the equilibrium constant provides information on the extent of the reaction. As a general guideline, if K_{eq} is less than 1, there are more reactants than products, i.e., the reaction does not proceed to a large extent. If K_{eq} is greater than 1, there are more products than reactants and the reaction proceeds significantly to the right.

- The **reaction quotient** takes the form of the equilibrium expression and describes the ratio of products to reactants at *any point in time*, not just at equilibrium. If the reaction in question is not at equilibrium, Q is not equal to K and may be greater than or less than K. As the reaction approaches equilibrium, Q approaches K and finally, at equilibrium, $Q = K_{eq}$.

- The amount of a solid that will dissolve in solution depends only on the relative strengths of the interionic forces compared to the attractive forces between water and the particular ions, not on the absolute amount of solid present. Once the solution is saturated, it is not possible to increase the concentration of the dissolved compound by adding more of the solid. Thus, the concentration of pure solids is treated as a constant. To simplify equilibrium expressions, this constant is usually combined with the equilibrium constant and therefore does not appear in the equilibrium expression. This combined constant is called **solubility product constant**, K_{sp}, for dissolution reactions of slightly soluble salts (**Explorations 4E, 4F**). The **molar solubility** (in mol/L), **solubility** (in g/L), and **solubility product constant** all provide information on the amount of an ionic substance that can dissolve in water.

Using Equilibrium Calculations

You should know how to:

- Predict equilibrium concentrations of slightly soluble salts using the equilibrium expression, the equilibrium constant, and knowledge of the stoichiometry of the reaction (**Exploration 4E and 4F**).

- Use K_{sp}, the molar solubility, and the solubility in g/L to describe the extent of a dissolution reaction (**Exploration 4E**).

- Use the reaction quotient, Q, to determine if a reaction has reached equilibrium or predict the direction in which a reaction will proceed (**Exploration 4F**).

- Relate the free energy, ΔG, of a reaction to the extent of the reaction, as determined by the reaction quotient, Q (**Exploration 4G**).

Thinking Skills

What general skills are you building for your resume?

As you worked through Session 4, you have also been developing some general problem-solving and scientific thinking skills that are valued in a wide range of professions and in academia. Here is a list of the skills that you have been building for your resume.

Data Analysis Skills

You should be able to:

- Plot a data set and interpret the plot to reach a conclusion about a system (**Explorations 4A, 4C, and 4G**).

Mathematics Skills

You should be able to:

- Rearrange algebraic expressions and use them to solve problems (**Explorations 4C, 4D, 4E, 4F, and 4G**).
- Use graphs to interpret data and make predictions (**Exploration 4G**).

Checking Your Progress

What progress have you made toward answering the Module Question?

Appendix 4A

Concentration Units for Equilibrium Expressions

For all of the equilibria we have examined so far, we have given concentrations in moles per liter, indicated by the square brackets, "[ ]", around a species. Although you may use any units that accurately represent a concentration, a standard set of units has been defined so equilibrium constants can be compared.

Aqueous Solutions

For dilute, aqueous solutions of dissolved substances, concentrations in moles of solute per liter of solution are the standard. For more concentrated solutions, concentrations are instead expressed as **activities** (a), defined for a substance B in solution, a_B, as

$$a_B = \gamma_b[B]$$

where γ_B is the **activity coefficient** with units of L/mol and a value that is experimentally determined for each substance. For dilute solutes, γ is equal to 1 L/mol and the activity is numerically equal to the molar concentration, but dimensionless. Activity coefficients take into account the fact that, at higher concentrations, solute molecules interact and may behave as if the concentration (in moles per liter) is higher or lower than it actually is.

Gases

For gases, the concentration in moles per liter is acceptable; however, the **partial pressure** of the gas can also be used as a concentration unit. Recall that the partial pressure of gas A is equal to the mole fraction (X_A) of that gas in the mixture times the total pressure of the system.

$$P_A = \left(\frac{n_A}{n_A + n_B + n_C + \ldots} \right) P_{total} = X_A P_{total}$$

These two different units (pressure in atmospheres and concentrations in moles per liter) will typically not give the same numerical value for the equilibrium constant. Thus, these equilibrium constants are distinguished by subscripts: K_C for concentration in moles per liter or K_P for partial pressure in atmospheres. For example, for the reaction of $SO_2(g)$ with $O_2(g)$ to give $SO_3(g)$:

$$2\,SO_2\,(g) \ + \ O_2\,(g) \rightleftharpoons 2\,SO_3\,(g)$$

The K_P equilibrium expression is written

$$K_P = \frac{P_{SO_3}^2}{P_{SO_2}^2 \cdot P_{O_2}}$$

The numerical value of K_P is related to K_C through the ideal gas law, using the following equation:

$$K_P = K_C(RT)^{\Delta n} = K_C(0.0821T)^{\Delta n}$$

where R is the gas constant, T is the temperature in Kelvin, and Δn is the change in the number of moles of gaseous components in the system as the reaction proceeds from products to reactants.

Mixed Phase Equilibria

Equilibria can be defined in any way that makes sense for a particular application, which sometimes results in the use of mixed units in the expression. When considering the solubility of gases in a solvent, **Henry's Law** defines the equilibrium between gas above a liquid and gas dissolved in a liquid, for example, in describing the dissolution of oxygen in water.

$$O_2\ (g) \underset{}{\overset{K_H}{\rightleftharpoons}} O_2\ (aq)$$

The equilibrium expression for the Henry's Law constant mixes units of gas pressure in atmospheres and solution concentration in moles per liter, with the equilibrium constant designated as K_H.

$$K_H = \frac{P_{O_2}}{\left[O_2(aq)\right]}$$

Units for Equilibrium Constants

You might be wondering why all of the equilibrium constants we have shown are unitless. How can this be, since the concentrations of the products and reactants have units associated with them? The answer is that we are assuming that concentrations in moles per liter or pressure in atmospheres are approximately equivalent to activities. Because activities are dimensionless, so too are equilibrium constants. Under most conditions that we will encounter, this practice is sufficient to obtain correct answers for equilibrium calculations. However, if very precise concentrations are needed or with concentrated solutions or at high pressures, the use of activities is essential.

Pure Solids and Liquids

Many equilibrium reactions contain pure solids or liquids as a product or reactant. The equilibrium expressions for reactions involving pure liquids and solids can be greatly simplified because the activity of a pure liquid or solid is, by definition, equal to one. This is effectively equivalent to assuming that the concentration of a pure liquid or solid is a constant at a given temperature and pressure. This can be demonstrated with some algebra and a knowledge of the density of the substance and its molecular weight.

For example, consider water as a pure liquid with a density of 1.00 g/mL at 25°C and a molecular weight of 18.0 g/mol. The concentration of water in moles per liter is:

$$[H_2O] = \frac{1.00\ g}{mL} \times \frac{1\ mol}{18\ g} \times \frac{1000\ mL}{L} = 55.5\ M$$

In dilute solutions where water is the main component, this concentration does not vary with the amount of water present but is a constant at a given temperature. The same logic holds for pure solids because they too have a characteristic density and molecular weight. The effect of this simplification on the writing of the equilibrium expression is that the constant concentration of the pure liquid or solid is incorporated into the equilibrium constant for a particular reaction.

Appendix 4B

Solubility Product Constants (K_{sp}) at 25°C

Compound	K_{sp}	Compound	K_{sp}
Aluminum		**Copper**	
$Al(OH)_3$	2×10^{-32}	$Cu_3(AsO_4)_2$	7.6×10^{-76}
Barium		$CuBr$	5.2×10^{-9}
$Ba_3(AsO_4)_2$	7.7×10^{-51}	$CuCl$	1.2×10^{-6}
$BaCO_3$	8.1×10^{-9}	CuI	5.1×10^{-12}
$BaCrO_4$	2.4×10^{-10}	$Cu(IO_3)_2$	7.4×10^{-8}
BaF_2	1.7×10^{-6}	Cu_2S	2×10^{-47}
$Ba(IO_3)_2 \cdot 2H_2O$	1.5×10^{-9}	CuS	9×10^{-36}
$BaC_2O_4 \cdot H_2O$	2.3×10^{-8}	$Cu(SCN)_2$	4.8×10^{-15}
$BaSO_4$	1.08×10^{-10}	$Cu(OH)_2$	2.2×10^{-20}
Cadmium		**Iron**	
$Cd_3(AsO_4)_2$	2.2×10^{-33}	$FeAsO_4$	5.7×10^{-21}
$Cd(OH)_2$	5.9×10^{-15}	$FeCO_3$	3.5×10^{-11}
$CdCO_3$	1.8×10^{-14}	$Fe(OH)_2$	8×10^{-16}
CdS	7.8×10^{-27}	$Fe(OH)_3$	4×10^{-38}
Calcium		$FePO_4$	1.2×10^{-18}
$Ca_3(AsO_4)_2$	6.8×10^{-19}	$Fe_3(PO_4)_2$	1×10^{-33}
$CaCO_3$	8.7×10^{-9}	FeS	5.0×10^{-18}
CaF_2	4.0×10^{-11}	**Lead**	
$Ca(OH)_2$	5.5×10^{-6}	$Pb_3(AsO_4)_2$	4.1×10^{-36}
$Ca(IO_3)_2 \cdot 6H_2O$	6.4×10^{-7}	$PbBr_2$	3.9×10^{-5}
$CaC_2O_4 \cdot H_2O$	2.6×10^{-9}	$PbCO_3$	3.3×10^{-14}
$Ca_3(PO_4)_2$	2.0×10^{-29}	$PbCl_2$	1.6×10^{-5}
$CaSO_4$	1.9×10^{-4}	$PbCrO_4$	1.8×10^{-14}
Chromium		PbF_2	3.7×10^{-8}
$Cr(OH)_2$	1.0×10^{-17}	$Pb(IO_3)_2$	2.6×10^{-13}
$Cr(OH)_3$	6×10^{-31}	PbI_2	7.1×10^{-9}
Cobalt		$Pb(OH)_2$	1.2×10^{-15}
$Co(OH)_2$	2×10^{-16}	$PbSO_4$	1.6×10^{-8}
$Co(OH)_3$	1×10^{-43}	PbS	8×10^{-28}

Compound	K_{sp}	Compound	K_{sp}
Manganese		Ag_2SO_4	1.6×10^{-5}
$Mn(OH)_2$	1.9×10^{-13}	Ag_2S	2×10^{-49}
$MnCO_3$	1.8×10^{-11}	**Strontium**	
Magnesium		$SrCO_3$	1.1×10^{-10}
$Mg(OH)_2$	1.2×10^{-11}	$SrCrO_4$	3.6×10^{-5}
$MgCO_3$	6.8×10^{-6}	SrF_2	2.8×10^{-9}
$MgC_2O_4 \cdot 2\ H_2O$	4.8×10^{-6}	$Sr(IO_3)_2$	3.3×10^{-7}
MgF_2	7.4×10^{-11}	$SrC_2O_4 \cdot H_2O$	1.6×10^{-7}
$Mg_3(PO_4)_3$	9.9×10^{-25}	$SrSO_4$	3.8×10^{-7}
Mercury		**Thallium**	
Hg_2Br_2	5.8×10^{-23}	$TlBrO_3$	8.5×10^{-5}
Hg_2Cl_2	1.3×10^{-18}	$TlBr$	3.4×10^{-6}
Hg_2I_2	4.5×10^{-29}	$TlCl$	1.7×10^{-4}
Hg_2SO_4	7.4×10^{-7}	Tl_2CrO_4	9.8×10^{-13}
HgS	4×10^{-53}	$TlIO_3$	3.1×10^{-6}
$Hg_2(SCN)_2$	3.0×10^{-20}	TlI	6.5×10^{-8}
Nickel		Tl_2S	5×10^{-21}
$Ni_3(AsO_4)_2$	3.1×10^{-26}	**Tin**	
$NiCO_3$	6.6×10^{-9}	SnS	1×10^{-25}
$Ni(OH)_2$	6.5×10^{-18}	$Sn(OH)_2$	1.4×10^{-28}
NiS	3×10^{-19}	**Titanium**	
Silver		$Ti(OH)_4$	1×10^{-40}
Ag_3AsO_4	1×10^{-22}	**Zinc**	
$AgBrO_3$	5.77×10^{-5}	$Zn_3(AsO_4)_2$	1.3×10^{-28}
$AgBr$	5.25×10^{-13}	$ZnCO_3$	1.4×10^{-11}
Ag_2CO_3	8.1×10^{-12}	$Zn_2Fe(CN)_6$	4.1×10^{-16}
$AgCl$	1.78×10^{-10}	$Zn(OH)_2$	1.2×10^{-17}
Ag_2CrO_4	2.45×10^{-12}	$ZnC_2O_4 \cdot 2H_2O$	2.8×10^{-8}
$AgIO_3$	3.02×10^{-8}	$Zn_3(PO_4)_2$	9.1×10^{-33}
AgI	8.31×10^{-17}	ZnS	1×10^{-21}
$Ag_2C_2O_4$	3.5×10^{-11}		
Ag_2O	2.6×10^{-8}		
Ag_3PO_4	1.3×10^{-20}		

Le Chatelier's Principle

How can we remove contaminants from a water supply?

Exploration 5A: The Storyline

Creating the Context

Why is this an important question?

In previous Sessions, you explored the processes by which contaminants find their way into a water supply. Water dissolves slightly soluble minerals from rocks, industry and agriculture contribute substances to waterways, and gaseous components in the atmosphere can "rain out" and end up in a water supply. In this Session, you will begin to think like a water treatment engineer to plan ways to *remove* the substances that are over the allowable limits for drinking water. The process of removing contaminants is called **remediation**.

Developing Ideas

Your instructor will demonstrate several potentially useful processes for removing contaminants from water. Your goal is to observe these methods, compare them, and think about how they work.

1. Jot down your observations as your instructor removes the contaminants from samples containing:

    a. metal ions, such as copper or iron

    b. colored organic compounds, such as indicator dyes

    c. high concentrations of Ca^{2+} and Mg^{2+}

2. For each sample, describe in your own words what happened to the contaminant. Where did it go?

3. Some of these processes will now be reversed. In general terms, list the methods used to reverse the cleanup processes.

Looking Ahead

The other Explorations in this Session can guide you to explore additional aspects of contaminant removal. These will be used at the discretion of your instructor.

*   **Exploration 5B:** How can we drive an equilibrium reaction to one side?

*   **Exploration 5C:** Which precipitating reagent will remove the most contaminant?

*   **Exploration 5D:** How much precipitating reagent is required for effective water treatment?

Exploration 5B

How can we drive an equilibrium reaction to one side?

Creating the Context

Why is this an important question?

What do you already know?

How can you find out more?

As you observed in Exploration 5A, some water contaminants can be removed by the processes of **precipitation**, where a solid is formed from aqueous ions, and **adsorption**, where the contaminant ion sticks to the surface of an insoluble reagent. We will take a closer look at these processes in Explorations 5C and 5D. Both of these reactions are equilibria that may or may not proceed to completion. In this Exploration, our goal is to see *How we can drive an equilibrium reaction to one side?*

Think back to Exploration 4C, where you examined a system approaching equilibrium. You saw that collisions between reacting particles (molecules or ions) are required for the reaction to proceed in either the forward or the reverse direction. Thus the rate of a reaction depends on the concentration of the species involved in the reaction and the temperature. For the generic reaction that follows,

$$aA \ + \ bB \ \underset{k_r}{\overset{k_f}{\rightleftharpoons}} \ cC \ + \ dD$$

the rates of the forward and reverse reactions can be expressed as a function of concentration at a particular temperature by the following equation:

$$\text{Rate (forward rxn.)} = k_f [A]^a[B]^b$$

$$\text{Rate (reverse rxn.)} = k_r [C]^c [D]^d$$

When a reaction begins and only reactants are present, collisions occur between reactants to produce products. Because the concentration of reactants is large, the number of collisions involving reactants is also large, and the rate of the forward reaction is fast. When the concentration of products builds up in the system, collisions involving product molecules or ions can occur and the rate of the reverse reaction begins to be significant. Thinking about the reaction at the microscopic level should give you some ideas on how to drive the reaction in a particular direction.

Our hypothesis in this Exploration will be based on **Le Chatelier's Principle**, which states that any system that is perturbed from an equilibrium position by a change in conditions (concentration, temperature, or pressure) will change in such a way that the system will move towards a new equilibrium position to compensate for the perturbation. You will change the concentrations of reactants and products in some reactions and observe the resulting changes in the equilibrium positions. Finally, you will relate your observations to the microscopic view of the reactions and explain them in terms of the rates of the forward and reverse reactions. By the end of the Exploration, you will know how changes in concentration affect the equilibrium position of a reaction.

Developing Ideas

In this exercise,[1] you will observe two equilibrium reactions between differently colored species so you can easily discern any shifts in the position of the equilibrium. Work in groups of three for this exercise.

PART I: IRON-THIOCYANATE EQUILIBRIUM

The first equilibrium reaction you will study is the reaction of ferric ions (Fe^{3+}) with thiocyanate ions (SCN^-) to form an iron-thiocyanate complex, $[Fe(SCN)]^{2+}$.

$$Fe^{3+} (aq) \; + \; SCN^- (aq) \; \rightleftharpoons \; [Fe(SCN)]^{2+} (aq)$$

Getting Started

1. Mix 2 mL of 0.01 M ferric chloride, $FeCl_3$, with 2 mL of 0.01 M potassium thiocyanate, KSCN in a 150 mL beaker. Note the colors of the two original solutions and the solution after mixing.

Solution	Color
$FeCl_3$	
KSCN	
Mixture	

   Answer the following questions in your laboratory notebook.

   a. What ions are available for reaction when the solutions are mixed?

   b. K^+, Cl^-, and SCN^- ions are colorless. What colors are Fe^{3+}(aq) and $[Fe(SCN)]^{2+}$(aq) ions?

Ion	Color
Fe^{3+} (aq)	
$[Fe(SCN)]^{2+}$ (aq)	

   c. Using your knowledge of reactant and product colors, what color change indicates that the concentration of $[Fe(SCN)]^{2+}$ (aq) has increased (i.e., the equilibrium has shifted to the right)?

   d. What color change indicates that the concentration of $[Fe(SCN)]^{2+}$ (aq) has decreased (i.e., the equilibrium has shifted to the left)?

2. Dilute the solution prepared in step **1** by adding 20 mL of distilled water to the beaker. Mix well and fill four test tubes with 5 mL of the diluted solution. Set one test tube aside for a color reference.

---

1. Adapted from an experiment developed by faculty at Canada College, Redwood City, CA.

Adding KCl

1. Take one of the test tubes and label it "+ KCl".

2. Add a crystal of KCl to the test tube and note your observations. Does the color of the solution change?

3. Answer the following questions in your laboratory notebook.

   a. Write the equation for the dissolution of KCl in water. What ions are available for reaction?

   b. Do these ions participate in the equilibrium you are studying?

   c. Did the position of the equilibrium shift noticeably when you added KCl to the solution? If so, how?

Adding FeCl$_3$

1. Take one of the test tubes and label it "+ FeCl$_3$".

2. Add a crystal of FeCl$_3$ to the test tube and note your observations. Does the color of the solution change?

3. Answer the following questions in your laboratory notebook.

   a. Write the equation for the dissolution of FeCl$_3$ in water. What ions are available for reaction?

   b. What happens to the rate of the forward reaction when you add FeCl$_3$? Explain in the context of collisions of particles.

   c. When FeCl$_3$ is first added, which rate is faster, the forward or the reverse? Explain in the context of a microscopic view and collisions of particles.

   d. How will the concentrations of Fe^{3+}, CNS^-, and $[Fe(SCN)]^{2+}$ change as time passes? What color changes support your statements?

   e. Once the reaction reaches an equilibrium condition again, how will the final concentration of $[Fe(SCN)]^{2+}$ compare to its initial concentration before FeCl$_3$ was added? Explain in the context of collisions of particles.

   f. Adding FeCl$_3$ resulted in the shift of the equilibrium in which direction, left or right?

Adding KSCN

1. Take one of the test tubes and label it "+ KSCN".

2. Add a crystal of KSCN to the test tube and note your observations. Does the color of the solution change?

3. Answer the following questions in your laboratory notebook.

   a. Write the equation for the dissolution of KSCN in water. What ions are available for reaction?

   b. What is the initial effect of adding KSCN on the rate of the reverse reaction?

   c. Which rate is faster, the forward or reverse? Explain in the context of collisions of particles.

   d. How will the concentrations of Fe^{3+}, SCN^-, and $[Fe(SCN)]^{2+}$ change as time passes? What color changes support your statements?

   e. Once the reaction reaches an equilibrium condition again, how will the final concentration of $[Fe(SCN)]^{2+}$ compare to its initial concentration before KSCN was added? Explain in the context of a microscopic view.

   f. Adding KSCN resulted in the shift of the equilibrium in which direction, left or right?

PART II: CHROMATE-DICHROMATE EQUILIBRIUM

The next equilibrium reaction you will study is the reaction of acid (H_3O^+) with yellow chromate ions (CrO_4^{2-}) to form orange dichromate ions ($Cr_2O_7^{2-}$) and water.

$$2\ H_3O^+ (aq)\ +\ 2\ CrO_4^{\ 2-} (aq)\ \rightleftharpoons Cr_2O_7^{2-} (aq)\ +\ 3\ H_2O\ (l)$$

$$\quad\text{colorless}\qquad\qquad\text{yellow}\qquad\qquad\qquad\text{orange}\qquad\quad\text{colorless}$$

1. Prepare two test tubes containing 2 mL of 0.1 M potassium chromate, K_2CrO_4, in each one. Set one test tube aside for a color reference.

2. To the other test tube, add 1-2 drops of 12 M hydrochloric acid, HCl. Mix well and note any color changes in your laboratory notebook. (*Tip:* HCl dissolves in water to form H_3O^+ ions and Cl^- ions.)

3. To the same test tube, add 2-3 drops of 12 M sodium hydroxide, NaOH. Mix well and note any color changes in your laboratory notebook. (*Tip:* OH^- ions react with H_3O^+ ions to form H_2O.)

4. To the same test tube, add 12 M HCl dropwise until a definite orange color persists. Now add 0.5 g of sodium borate, Na_3BO_3. Mix well and note any color changes in your laboratory notebook. (*Tip:* Neither Na^+ or BO_3^{3-} ions react with chromate, dichromate, or K^+ ions when added to the reaction mixture.)

5. Explain the color changes you observed in steps 1–4 in terms of shifting the equilibrium position of the reaction.

Working with the Ideas

The following questions will help you generalize your results from the experiments.

6. Which way does an equilibrium shift when you:
    a. *add reactants* to a reaction at equilibrium?
    b. *add products* to a reaction at equilibrium?
    c. *remove reactants* from a reaction at equilibrium?
    d. *remove products* from a reaction at equilibrium?

7. For the equilibrium shown below, which of the following would reduce the amount of lead (Pb^{2+}) in solution?

$$Pb^{2+} (aq)\ +\ CO_3^{\ 2-} (aq) \rightleftharpoons PbCO_3\ (s)$$

    a. Adding a source of CO_3^{2-} ions
    b. Adding a source of Pb^{2+} ions
    c. Removing CO_3^{2-} ions
    d. Removing some of the $PbCO_3$ solid
    e. Adding more $PbCO_3$ solid

8. How might you use the principles you developed in this Exploration to remove contaminants from a water supply?

Exploration 5C

Which precipitating reagent will remove the most contaminant?

Creating the Context

Why is this an important question?

Precipitation reactions are commonly used to remove unwanted contaminants from drinking water. Many of those substances made their way into the water supply through dissolution of minerals. Water treatment processes drive the dissolution reaction in the reverse direction to precipitate the impurity out of solution. Under conditions in which equilibrium can be achieved (a closed system with enough reactants and products present to supply equilibrium concentrations), both precipitation and dissolution can be treated as equilibrium reactions, and we can use our knowledge of solubility equilibria to answer the question *Which precipitating reagent will remove the most contaminant?*

Consider a water supply with too much calcium, Ca^{2+}. One way to remove the Ca^{2+} ions is to add a source of an anion that will react to form an insoluble solid with Ca^{2+}. The table of solubility product constants in Appendix 4B on page 97 gives a number of solubility equilibrium reactions in which calcium participates.

$$CaX\ (s) \rightleftharpoons Ca^{2+}\ (aq)\ +\ X^{2-}\ (aq)$$

or

$$CaX_2\ (s) \rightleftharpoons Ca^{2+}\ (aq)\ +\ 2\ X^-\ (aq)$$

Ions that form slightly soluble precipitates with calcium include CO_3^{2-}, SO_4^{2-}, OH^-, PO_4^{3-}, F^-, and IO_3^-. Recall that by convention, solubility equilibria are written as dissolution reactions, with the K_{sp} values corresponding to the reaction written as a dissolution. As an example, consider the precipitation of calcium as the carbonate salt, $CaCO_3$. If enough CO_3^{2-} is added to the solution, Q will be greater than K_{sp} and precipitation will occur.

An easy way to think about such precipitation reactions is to break them into two artificially separated steps.

- **Stoichiometric precipitation:** The contaminant ion reacts completely with the precipitating ion, according to the stoichiometry of the reaction, in this case, one mole of calcium per mole of carbonate. In the process, a stoichiometric amount of the added precipitating ion is used up.

$$Ca^{2+}\ (aq)\ +\ CO_3^{\ 2-}\ (aq) \xrightarrow{K_{eq}} CaCO_3\ (s)$$

The equilibrium constant for this reaction is the *inverse* of the K_{sp} for the dissolution reaction. Because this equilibrium constant is typically very large for the precipitation of slightly soluble salts, we can view the reaction as having proceeded nearly to completion.

- **Dissolution:** We can determine the residual concentration of the ions by considering the re-dissolution of the precipitated solid to establish equilibrium.

$$CaCO_3\ (s) \xrightarrow{K_{sp}} Ca^{2+}\ (aq)\ +\ CO_3^{2-}\ (aq)$$

What background information is helpful?

The equilibrium constant for this reaction is simply K_{sp}, the solubility product constant. The concentration of the dissolved species can be calculated using K_{sp}, just as if you had made a saturated solution by dissolving $CaCO_3$ in water.

In reality, these two steps are occurring simultaneously, but separating them into two clearly defined reactions makes the calculations easier to conceptualize.

To understand precipitation reactions better, you will now view an animation of a precipitation reaction and think about the process on a microscopic level. You will then think about how to choose a reagent that will remove the maximum amount of a contaminant ion.

How can you find out more?

Developing Ideas

The following activities and questions are based on chemical reactions in which equivalent molar amounts (stoichiometric amounts) of a reagent are added to precipitate contaminant ions.

1. The removal of metal ions from a water supply is often done by precipitating the ions out of solution. In this exercise, we will examine an animation of the removal of copper ions by precipitation with iodate, IO_3^-. You may find the animation at http://chemistry.beloit.edu/Water/moviepages/PrecipitationCu.htm. The animation begins with the depiction of a solution of copper sulfate, $CuSO_4$. Sodium iodate, $NaIO_4$, is added as the precipitating reagent. The chemical equation that describes the reaction is

   $$Cu^{2+} (aq) \;+\; 2\, IO_3^- (aq) \;\rightleftharpoons\; Cu(IO_3)_2 (s)$$

   a. What must happen before precipitation can occur?

   b. Once the $Cu(IO_3)_2$ precipitates out of solution, what observations can you make about the stoichiometry of the solid in the bottom of the container?

   c. Write the equilibrium expression for the *precipitation* of $Cu(IO_3)_2$.

2. Your instructor will present an animation of the *dissolution* of copper iodate, $Cu(IO_3)_2$ to form a saturated solution. The chemical equation that describes the reaction is

   $$Cu(IO_3)_2 (s) \;\rightleftharpoons\; Cu^{2+} (aq) \;+\; 2\, IO_3^- (aq)$$

   a. Once the $Cu(IO_3)_2$ precipitate begins dissolving, what observations can you make about the stoichiometry of the solid in the bottom of the container?

   b. Write the equilibrium expression for the *dissolution* of $Cu(IO_3)_2$.

3. How are precipitation and dissolution similar in terms of the type and concentrations of dissolved ions present after the system has reached equilibrium? How are they different?

4. Assume that the precipitation reaction is carried out under stoichiometric conditions, i.e., for every mole of Cu^{2+} present, 2 moles of IO_3^- are added. Under these conditions, what are the molar concentrations of Cu^{2+} (aq) and IO_3^-(aq) once the solid has precipitated and the system has reached equilibrium? Copper iodate has a $K_{sp} = 7.4 \times 10^{-8}$.

5. What will be the equilibrium molar concentration of Cu^{2+} if you use a stoichiometric amount of the sulfide ion (S^{2-}) instead of iodate (IO_3^-) to precipitate Cu^{2+} out of solution ($K_{sp}(CuS) = 9 \times 10^{-36}$)? Which of the two anions will most effectively remove Cu^{2+} ions from the solution?

6. When you add a precipitating reagent to a solution, you can't *just* add an anion or a cation. Since all ionic compounds exist as combinations of anions and cations, you will need to add a *compound* to the solution that contains both the desired anion and a cation. Think back to Exploration 3A and the solubility rules you generated. If you wish to add a soluble compound containing an anion that precipitates the contaminant, which cations would you choose to accompany your anion?

Working with the Ideas

The following questions are based on stoichiometric reactions, in which equivalent molar amounts (stoichiometric amounts) of a reagent are added to precipitate contaminant ions.

1. Sulfide-producing bacteria are now being used experimentally in water treatment plants to remove heavy metals from wastewater. The sulfide ion (S^{2-}) reacts with metal ions to form a slightly soluble solid that adheres to the mass of bacteria and can be removed by settling or filtration. Consider a wastewater contaminated with nickel, Ni^{2+}. The solubility product constant, K_{sp}, for nickel sulfide (NiS) is 1.4×10^{-24}.

    a. Write the equilibrium expression and determine the value of the equilibrium constant for the precipitation reaction shown below.

    $$Ni^{2+} (aq) \ + \ S^{2-} (aq) \ \rightleftharpoons \ NiS (s)$$

    b. What is the final concentration of Ni^{2+} when this reaction has reached equilibrium, assuming equimolar concentrations of Ni^{2+} and S^{2-}?

    c. If the water has an initial concentration of Ni^{2+} equal to 0.01 M and you add a stoichiometric amount of sulfide, how many grams of NiS will precipitate per liter of water?

2. You are trying to purify a water supply that contains lead (Pb^{2+}) at a concentration of 200 µg/L in order to make it safe to drink. You know the following:

    •   The MCL of lead allowed by the Safe Drinking Water Act is 15 µg/L. Possibly useful conversion factors: 10^{6}µg/g, 10^{3}mg/g. Density of water = 1.00 g/mL.

    •   A number of slightly soluble lead salts will precipitate if the appropriate anion is added. These lead salts and their K_{sp} values are listed in Appendix 4B.

    a. Which of the possible anions would reduce the concentration of Pb^{2+} to the lowest level and is below the MCL? Show your work.

    b. As you read in the **Developing Ideas** section, you cannot add anions by themselves—cations must accompany them. What ionic compound(s) could you add to the water to deliver the appropriate amount of the anion for precipitating the lead ions without otherwise exceeding MCLs?

    c. If you add a stoichiometric amount of reagent, how much of the reagent you chose in **2b** will you need to add in grams of reagent per liter of water?

    d. Once the stoichiometric treatment is complete, how much lead (Pb^{2+}) will remain in solution? Give the answer in both moles/L and µg/L. What process is responsible for the lead in the solution at this point?

    e. What other factors must be considered about this purification procedure before you can use it for a drinking water supply?

3. You would like to remove the calcium from a water supply containing 324 mg of Ca^{2+} per liter by precipitating it from solution with CO_3^{2-}. You wish to determine how much sodium carbonate to add for maximum calcium removal.

a. If you add a stoichiometric amount of sodium carbonate (i.e., an equal mole quantity relative to the reactant species, Ca^{2+}), how many grams of Na_2CO_3 will you need to add per liter of water?

b. Once the $CaCO_3$ has precipitated and the solution has reached equilibrium, what is the final concentration of Ca^{2+} in the water in mol/L and in mg/L?

c. How many grams of $CaCO_3$ will precipitate per liter of water?

4. Sodium carbonate dissolves in water to produce an equilibrium mixture of car-

$$CO_3^{2-} \text{ (aq)} \quad + \quad H_2O \text{ (l)} \rightleftharpoons HCO_3^- \text{ (aq)} \quad + \quad HO^- \text{ (aq)}$$

bonate (CO_3^{2-}) and bicarbonate (HCO_3^-) ions, with $K_{eq} = 2.2 \times 10^{-4}$. The CO_3^{2-} ion can be favored by raising the concentration of OH^- and pushing the equilibrium towards the reactants. If you wish to ensure that 99% of the carbon-containing species present are in the form of CO_3^{2-}, what is the desired concentration of hydroxide?

5. Using the data in Appendix 4B on page 97, determine which anions will most effectively remove the following cations from the following wastewaters:

a. Ag^+ from a photographic processing plant

b. Ba^{2+} from medical waste

c. Ni^{2+} from a semiconductor processing plant

d. Cu^{2+} from mine drainage water

e. Which precipitating anions would you eliminate because of potential toxicity? See the MCLs on the Environmental Protection Agency's web site at http://www.epa.gov/waterscience/drinking/.

Exploration 5D

How much precipitating reagent is required for effective water treatment?

Creating the Context

Why is this an important question?

After working through the precipitation reactions in Exploration 5C, you may think that you have learned all that is necessary to purify a water supply. After all, these precipitation reactions satisfied the first goal of any water treatment engineer: that the water be safe to drink after treatment. But water treatment engineers also need to take into account the *time* required for treatment and the *cost* of the treatment process. These two factors are intimately related, and there are always trade-offs between the two.

For example, many of the precipitation reactions used in water treatment are quite slow, taking up to 24 hours to reach equilibrium. Whether the community decides to allow the water to sit for 24 hours is related to cost and demand for the water. A preliminary list of questions the engineers will need to ask follows:

• How much water does the community typically use?

• How many treatment tanks can they afford to build? and to maintain?

• How much land is available to build them on?

- Is it necessary for the precipitation reaction to reach equilibrium for the concentrations of contaminants to be lowered below the MCLs?
- What will happen to the water pipes if a precipitation reaction continues to occur in the distribution pipes?
- Can they speed up the precipitation process so the water doesn't have to be held for such a long time in treatment tanks?

What background information is helpful?

The answers to these questions may differ, according to the needs of the communities involved. An urban water treatment plant may not have much land to build on, or the land may be very expensive, whereas a rural water treatment plant may have ready access to inexpensive land. Some problems can be minimized by changing the *process* of water treatment instead of the facilities.

So far we have considered only stoichiometric precipitation reactions, in which equivalent molar amounts of a reagent are added to precipitate contaminant ions. In this Exploration, you will use an *excess* of reagent to precipitate a contaminant ion. The shift in the equilibrium position of a solubility reaction by adding an excess of one of the ions of the compound is called the **common ion effect;** it is an application of Le Chatelier's principle. The solubility equilibrium is shifted to the left by adding an ion that precipitates the contaminant ion. This process costs more because more reagents must be used, but may allow a water treatment plant to function without building more holding tanks.

Developing Ideas

Consider the water hardness ion Ca^{2+}, which can be removed by precipitation with CO_3^{2-}.

$$Ca^{2+} (aq) + CO_3^{2-} (aq) \rightleftharpoons CaCO_3 (s)$$

Figure 5-1 shows the relationship between the concentration of calcium and the concentration of carbonate in a solution that is an equilibrium mixture of the two ions.

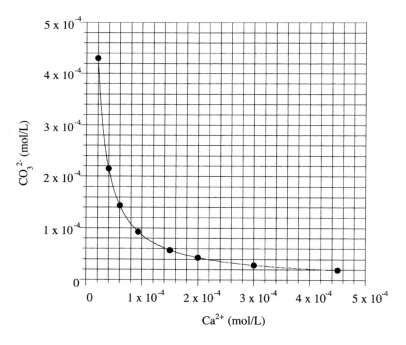

Figure 5-1: This plot shows the relationship of $[Ca^{2+}]$ to $[CO_3^{2-}]$ at equilibrium.

1. When pure $CaCO_3$ dissolves in water, equimolar amounts of Ca^{2+} and CO_3^{2-} go into solution. Plot the following points on the graph in Figure 5-1 and draw a line connecting them.

    *   0.0 mol/L of $CaCO_3$ dissolved
    * 0.25×10^{-4} mol/L of $CaCO_3$ dissolved
    * 0.5×10^{-4} mol/L of $CaCO_3$ dissolved
    * 0.75×10^{-4} mol/L of $CaCO_3$ dissolved

2. The curved line on the graph represents the concentrations of the two ions under equilibrium conditions.

    a. At what concentration of Ca^{2+} and CO_3^{2-} does the line you drew in problem 1 intersect the curved line? (*Hint:* Begin by writing the equilibrium expression. The K_{sp} for $CaCO_3$ is 8.7×10^{-9} at 25°C.)

    b. Choose a point in the region *below the curve* and estimate the x,y coordinates that give the values for $[Ca^{2+}]$ and $[CO_3]^{2-}$. Calculate the value of Q at this point and compare it to K_{sp}. Will precipitation occur if the concentrations are in the region below the curve?

    c. Choose a point in the region *above the curve* and estimate the x,y coordinates that give the values for $[Ca^{2+}]$ and $[CO_3]^{2-}$. Calculate the value of Q at this point and compare it to K_{sp}. Will precipitation occur if concentrations are in the region above the curve?

3. Consider a solution that is 3×10^{-4} M in Ca^{2+}. An amount of Na_2CO_3 is added to give an initial concentration of 3×10^{-4} M in CO_3^{2-}.

    a. Where is this point on the plot?

    b. What is the value of Q under these conditions?

    c. Will precipitation occur?

Working with the Ideas

The following problems will help you understand the concepts in more depth.

4. Consider a solution that is 4.2×10^{-3} M in Ca^{2+}. An amount of Na_2CO_3 is added that results in an initial concentration (before precipitation) of 4.2×10^{-3} M in CO_3^{2-}. Determine the equilibrium concentrations of Ca^{2+} and CO_3^{2-} after stoichiometric precipitation occurs. Enter your results in the following table.

	$CaCO_3$ (s) (mol/L)	Ca^{2+} (aq) (mol/L)	CO_3^{2-} (aq) (mol/L)	Q
Initial Conc.	constant			
Conc. after stoichiometric precipitation	constant			

5. Now consider the effect of adding a *greater than stoichiometric* amount of a reagent to the same 4.2×10^{-3} M solution of Ca^{2+} by adding an amount of solid Na_2CO_3 that results in a concentration of 5.3×10^{-3} M in CO_3^{2-}.

    a. According to Le Chatelier's principle, which direction would you expect the following reaction to shift when *excess* CO_3^{2-} is added?

$$CaCO_3 \text{ (s)} \underset{}{\overset{K_{sp}}{\rightleftharpoons}} Ca^{2+} \text{ (aq)} + CO_3^{\,2-} \text{ (aq)}$$

   b.   Assuming complete reaction, what amount of CO_3^{2-} is used up in the precipitation reaction?

   c.   What amount of CO_3^{2-} remains in solution unreacted?

   d.   Estimate the residual concentration of Ca^{2+} using Figure 5-1.

6.   After the precipitation reaction has gone to completion, some unreacted CO_3^{2-} *plus* some CO_3^{2-} from the dissolution of the solid $CaCO_3$ remain in solution.

   a.   Describe the equilibrium concentrations of Ca^{2+} and CO_3^{2-} under conditions of excess CO_3^{2-} *in terms of x*. Enter your results in the table below.

	$CaCO_3(s)$ (mol/L)	Ca^{2+} (aq) (mol/L)	CO_3^{2-} (aq) (mol/L)
Initial Conc. (after stoichiometric precipitation, assuming complete reaction)	constant		
Equilibrium Conc. (taking into account dissolution of the solid) in terms of x	constant		

   b.   Is x significant compared to the amount of excess CO_3^{2-} in solution?

   c.   Substitute the final equilibrium concentrations from the table into the equilibrium expression and solve for x to obtain the concentrations of both Ca^{2+} and CO_3^{2-} at equilibrium in moles per liter.

   d.   Has adding excess CO_3^{2-} achieved your goal of removing more Ca^{2+} than was removed with a stoichiometric precipitation? Explain.

7.   How does the *molar solubility* of $Mg(OH)_2$ change in a solution of 6.5×10^{-3} M hydroxide?

8.   A water treatment engineer is using $Mg(OH)_2(s)$ as a hydroxide source to precipitate heavy metal ions. He has added some $Mg(OH)_2$ solid, but the concentration of OH^- is still not high enough. He adds more $Mg(OH)_2$ solid, but the concentration of hydroxide doesn't change. Why not?

9.   Adding more than a stoichiometric amount of a precipitating reagent often speeds up reactions or removes more of a contaminant. When would it NOT be a good idea to add more than a stoichiometric amount of a treatment reagent? (*Hint:* Review the MCLs for various ions given in Session 1.)

Making the Link

Looking Back: What have you learned?

What chemistry experience have you gained?

Your experience in working with the chemical principles in Session 5 should have developed your skills to do the items on the following lists.

Le Chatelier's Principle

You should be familiar with:

- The concept that the position of an equilibrium reaction can be altered by changing the concentrations of products or reactants (**Explorations 5A** and **5B**).

- The idea that a reaction that is disturbed from equilibrium will always shift in a direction that allows a new equilibrium to be established (**Explorations 5A, 5B,** and **5C**).

- The common ion effect, where addition of an excess amount of one ion involved in a precipitation reaction shifts the solubility equilibrium to the left and reduces the concentration of the counterion to lower levels than could be achieved with no added ion (**Exploration 5D**).

Microscopic View

You should be able to:

- Discuss why Le Chatelier's Principle works in the context of changes in the rates of the forward and reverse reactions (**Exploration 5B**). The rate of the forward reaction is proportional to the amount of reactants, and the rate of the reverse reaction is proportional to the amount of products. Adding reactants or products changes the rates of the forward and reverse reactions, and the equilibrium position will shift. In all cases, a new equilibrium position is achieved when the rates of the forward and reverse reactions are again equal.

- Explain precipitation using a microscopic view. Collisions of the ions in solution lead to precipitation. The equilibrium is dynamic, and dissolution is also occurring simultaneously. At equilibrium, the rate of precipitation is equal to the rate of dissolution (**Exploration 5C**).

Using Equilibrium Calculations for Precipitation Reactions

You should know:

- How stoichiometric precipitation reactions work, where an equal molar quantity of a precipitating ion is added to remove a contaminant ion (**Exploration 5C**).

- How to use a table of K_{sp} values and the MCLs to select the optimum reagent for removing a contaminant ion (**Exploration 5C**).

- How to use the reaction quotient, Q, to determine whether precipitation will occur. If Q is less than K_{sp} (i.e., the concentrations of the participating ions aren't large enough to support an equilibrium condition), precipitation will not occur. Under these conditions, the reaction cannot reach equilibrium because the solution is not saturated and there is no solid present. If Q is greater than K_{sp}, precipitation will occur until the concentrations of the reactants and products have reached equilibrium values. Once the reaction has reached equilibrium, $Q = K_{sp}$ (**Exploration 5D**).

- How to remove more of a contaminant ion by adding an excess of precipitating reagent (**Exploration 5D**).

Thinking Skills

What general skills are you building for your resume?

What progress have you made toward answering the Module Question?

You have also been developing some general problem-solving and scientific thinking skills that are valued by employers in a wide range of professions and in academia. Here is a list of the skills that you have been building for your resume.

Mathematics Skills

You should be able to:

- Rearrange algebraic expressions and use them to solve problems (**Explorations 5C, 5D**).
- (Use graphs to predict and interpret results (**Exploration 5D**).

Checking Your Progress

Remediation

What procedures can you design to remove contaminants from a water supply?

Exploration 6A: The Storyline

At the beginning of the module, you laid out a plan for transforming a water source into a potable water supply. So far, you have learned what types of contaminants to expect in natural waters (Session 1) and why they are present (Session 3). You have also explored the concepts of equilibrium and how they explain why some water supplies contain more minerals than others (Session 4), and you have learned the theory behind using solubility equilibria to remove contaminants from water (Session 5). In Session 2, you learned several methods of analysis and determined the concentration of total dissolved solids (TDS), total alkalinity, and one of either fluoride, water hardness, or iron in an unknown sample.

The next step in implementing your water treatment plan is to apply some of the concepts you have learned to a real system in the laboratory. You will probably encounter some surprises in this laboratory work, since real life does not often act exactly as we've calculated that it should. In addition, there are many "right" ways to remove a contaminant, so do not feel constrained by the procedures your labmates are using. Be creative! The goal of this Session is to answer the question, *What procedures can you design to remove contaminants from a water supply?*

BATCH AND FLOW SYSTEMS FOR WATER TREATMENT

No matter what substance is used to remove contaminants from a sample, the goal is contaminant- and particle-free water. All substances that remove contaminants, in whatever way, are called *remediation agents*. Since all of the remediation agents are nearly insoluble solids or produce insoluble precipitates on reaction with contaminants in water, these solids have to be removed, or *filtered*, from the water after remediation.

Filtration of millions of gallons of water is not an easy task. Water treatment plants cannot use filter paper to remove solids because paper would clog, tear, or need replacement too often. Instead, water treatment plants filter undissolved solids by allowing the water to flow though a bed of sand or pulverized minerals. Sand is cheap, effective, and needs replacement only infrequently. Another advantage is that the flow rate through a bed of sand remains nearly constant for a long time.

Water treatment plants generally use one of two methods for remediation, the flow method or the batch method.

In the **flow method,** the remediation agent(s) is added directly to or on top of the bed of sand used for filtration. As the water flows through the system, remediation and filtration occur simultaneously, and a continuous stream of water can be treated (see Figure 6-1).

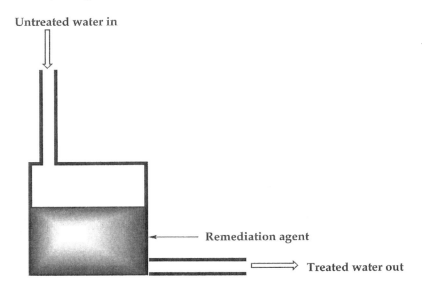

Figure 6-1:   Schematic of a flow system for water treatment.

In the **batch method**, the remediation agent(s) and water are added to a large reactor, then stirred to mix the reagents into the water. The solids are usually allowed to settle before filtration to reduce the load on the filtration system. The water is then filtered through a bed of sand to separate the water from any remaining solids (see Figure 6-2).

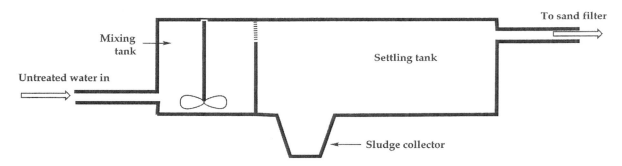

Figure 6-2:   Schematic of a batch reactor. Water is mixed for a set amount of time with water treatment chemicals in the mixing tank, then passed into the settling tank to allow most of the solids to settle out as sludge. After settling, the water is filtered through sand to remove any remaining particulates.

The chemical reactions used in water treatment processes proceed to equilibrium at a rate that depends on a number of factors, including temperature, concentrations of reacting species, and the surface area of the reagent used. By experimenting with these variables, engineers can "optimize" a treatment process: that is, make it occur within a reasonable amount of time using a minimum amount of chemicals. Another important variable is the **contact time**, or the amount of time the water is in contact with the remediation agent. In the flow method, this time is related to how long it takes for the water to flow through the system. In the batch method, the contact time is the time the water and remediation agent are in contact in the mixing and settling tanks before filtration.

ADJUSTING VARIABLES

Your goal in the next several laboratory periods is to develop a workable plan to reduce the concentration of a contaminant in a water sample to below the MCL. You can experiment with many variables to optimize your water treatment plan for effectiveness, convenience, and lowest cost. A brief description of each variable follows.

Batch vs. Flow System

Batch and flow system treatment plans each have advantages and disadvantages. A flow system typically requires less space because it is a one-step process and eliminates the need for holding basins; it may therefore be cheaper where land is expensive. On the other hand, flow systems clog more easily than batch reactors and require more stringent maintenance. Batch systems are best for treatment procedures that take some time to reach equilibrium and drive the concentration of the contaminant below the MCL. While it is usually possible to speed up slow reactions by adding more reagent, the cost can be prohibitive.

Type of Reagent

Usually more than one reagent will effectively remove the target ion. Each reagent has advantages and disadvantages in terms of cost, ease of use, and speed of the treatment process under the conditions used.

Amount of Reagent

Increasing the amount of reagent can speed up a remediation process, but it will also make it more expensive. Conversely, if the process works with the amount you initially chose to use, you may be able to use less reagent and save some money. Even a small increase or decrease in the amount of reagent(s) used will result in a dramatic cost difference when millions of gallons are being treated.

Contact Time

It doesn't matter how cheap a remediation scheme is if it can only remediate a few thousand liters an hour. Usually, the flow system has the advantage here.

- In your laboratory mimic of a flow system water treatment plant, you can vary the contact time using the plunger of the syringe (Appendix 6A). To prolong the contact time, insert the plunger into the syringe, leaving it at the very top of the syringe: the flow through the column will slow down and then stop. Alternatively, you can mix the reagent evenly with the sand so the solution is exposed to the reagent during its entire journey through the sand. To decrease the contact time, depress the plunger to apply pressure to the system and force the liquid to flow out at a faster rate.
- In a batch treatment system, the contact time is varied by allowing the sample to stir with the reagent for different lengths of time.

Surface Area of the Reagent

Many water treatment processes rely on the reaction of the surface of a solid reagent with a dissolved contaminant in the water. An increase in the surface area of the reagent can therefore increase the effectiveness of the treatment. Larger surface area can be achieved by using reagents with a smaller particle size, which results in a greater surface area per unit mass of reagent. The drawback is that a smaller particle size can reduce the flow rate of water through the filtration system. The batch method has a definite advantage over the flow system in this regard, because all of the sample is in contact with the remediating agent, and it is constantly agitated. This takes more time, but it is more effective for slower-acting remediation agents.

Temperature

The speed and extent of some of the water treatment reactions are dramatically affected by temperature. When using the batch method, it is easy to heat or cool the beaker containing the sample to optimize the treatment process. It is more difficult to control the temperature of a flow system. Although a change in temperature may decrease the amount of time and reagent needed to remediate the sample, watch out! Heating or cooling water is more expensive than you might think.

Developing Ideas

Before you begin your laboratory work, meet with the other students in your laboratory group; all of you will have the same contaminant. A wide variety of experiments are possible. Try to divide the work to maximize your efficiency in exploring different possibilities. Proceed as follows:

1. Read the section, **Preparing for Inquiry**, in this Exploration and the Exploration in this Session that is specific to the ion you are trying to remove. You'll need the information they contain for the next steps.

    *   Fluoride: **Exploration 6B**, page 119
    *   Hardness: **Exploration 6C**, page 122
    *   Iron: **Exploration 6D**, page 130

2. Working with your group and keeping the list of variables in mind (see **Preparing for Inquiry**), decide on an area for each person to investigate. The reagents available to remediate your specific ion can be found in the relevant Exploration.

3. Decide which variables will be changed and which will be held constant. Remember that you will learn NOTHING if you change too many variables at once!

4. In your laboratory notebook, sketch out a brief procedure for at least the *first three* experiments you plan to try. You may wish to create a table in your notebook like that shown below to keep track of your experiments and the results.

Trial	Reagent used	Amount of reagent (mg)	Contact time (minutes)	Ion conc. (mg/L)
untreated sample	------	------	------	
#1				
#2				
#3				
.				
.				

When trying "free-form" experiments in the laboratory, it is *critical* to plan your work carefully and take copious notes on the results. It is very easy to forget what you did if you don't write it down. When in doubt, write it down!

5. Jot down in your lab notebook which method you think will be most effective for removing the contaminant you've been assigned, and why. It may be interesting to compare your results with your predictions!

6. Your instructor may request that you fill in the planning worksheet on the following page to hand in.

Looking Ahead

Your laboratory group will be assigned one of the following Explorations. They contain information on the contaminant you will be trying to remove.

*   **Exploration 6B:** How can excess fluoride be removed from a water supply?
*   **Exploration 6C:** How can excess calcium and magnesium be removed from a water supply?
*   **Exploration 6D:** How can excess iron be removed from a water supply?

EXPLORATION 6A: WATER TREATMENT GROUP PLANNING WORKSHEET

Your name: _____   **Date:** _____

Names of other people in the group: _____

_____

Contaminant ion: _____

Sample Number: _____

Which treatment method(s) will you investigate? (Batch or flow, which reagent?)

Which variables will be changed and which held constant?

Experiment #1:

Experiment #2:

Experiment #3:

Exploration 6B

How can excess fluoride be removed from a water supply?

Creating the Context

Why is this an important question?

What do you already know?

In Exploration 2B, you saw that excess fluoride in water can be a health hazard, causing damage to bones and teeth and, at very high levels, acute toxicity that can be fatal. Thus, communities with high concentrations of fluoride in their water must remove it to levels below the MCL.

In Session 5, you learned about precipitation reactions that remove soluble contaminants from a water supply. Precipitation reactions require the addition of a precipitating agent and filtration of the resulting precipitate. **Adsorption reactions** require the addition of an insoluble solid with high surface area that the contaminants adhere to, followed by filtration to remove the insoluble solids. In general, the greater the surface area of the solid (finer particle size = more surface area), the more sites the substance can adhere to. There are trade-offs, however. The finer the particle size, the slower the water flow through a flow system.

Several processes can remove excess fluoride from water supplies.:

- Precipitation of calcium fluoride by adding calcium salts
- Adsorption to alumina, Al_2O_3
- Adsorption to activated carbon

In this Exploration, you will use the plan you developed in Exploration 6A to test different methods for removal of fluoride from your water sample. After experimenting with variables for several laboratory periods, you should be able to answer the Exploration Question *How can excess fluoride be removed from a water supply?*

Developing Ideas

1. One way to remove fluoride is by stoichiometric precipitation with calcium to form the slightly soluble salt calcium fluoride, CaF_2 (see Exploration 5C to review the concept of stoichiometric precipitation). The following exercises will help you determine which precipitating reagent to use and how much of it is needed for a stoichiometric precipitation. Remember that you can remove *more* fluoride if you add more calcium (common ion effect, see Exploration 5D), but it will cost more and may not be necessary to remove the required amount of fluoride.

   a. Begin by writing the equation for the precipitation reaction of CaF_2. What is the mole ratio of Ca^{2+} to F^- in the reaction?

   b. Using your results from Session 2B, where you measured the concentration of F^- in your sample in mg/L, calculate the concentration of F^- in your sample in moles/L.

   c. Calculate the stoichiometric amount of Ca^{2+} (in mol) required to remove F^- from 10 mL of your sample, assuming that the precipitation reaction goes to completion.

   d. Choose a calcium source that will provide Ca^{2+} with the least amount of other contaminants added. You will need to check the MCLs on the Environmental Protection Agency's web site at: http://www.epa.gov/waterscience/drinking/

Available sources of calcium for this experiment are
$Ca(NO_3)_2$, $CaCl_2$, $CaSO_4$, $CaCO_3$, and $Ca(OH)_2$

e. Calcium chloride ($CaCl_2$) and calcium nitrate ($Ca(NO_3)_2$) are both available as 0.005 M solutions. Calculate the volume of these solutions (in drops) required to precipitate the fluoride ion stoichiometrically. There are 20 drops per mL.

f. The other calcium sources ($CaSO_4$, $CaCO_3$, $Ca(OH)_2$) are available as slightly soluble solids. Look up the K_{sp} values for *one* of these compounds in Appendix 4A and determine the concentration of Ca^{2+} in a saturated solution of that compound. Calculate the volume of this saturated solution (in drops) required to precipitate the fluoride ion stoichiometrically. There are 20 drops per mL.

2. As you learned in Session 4, the assumption that a chemical reaction goes to completion is not necessarily correct. Many reactions reach an equilibrium position where measurable amounts of both products and reactants are present. Precipitation reactions of slightly soluble salts such as CaF_2 fall in this category; thus, if you add a stoichiometric amount of Ca^{2+} intending to precipitate out all of the fluoride in your sample, you will not actually remove *all* of the fluoride because the precipitation reaction does not go to completion.

a. Write the solubility product expression for CaF_2. The K_{sp} for CaF_2 is 4.0×10^{-11}. Will adding a stoichiometric amount of Ca^{2+} to *your* sample remove enough fluoride so the concentration is below the Maximum Contaminant Level? (See MCLs on the Environmental Protection Agency's web site at: http://www.epa.gov/waterscience/drinking/.)

b. Calculate the residual calcium remaining in the solution after stoichiometric precipitation of fluoride. Will this amount of calcium exceed acceptable levels of water hardness? See Table 2-1 on page 32.

c. Calculate the concentration of the ion that accompanies your calcium source, e.g., chloride for $CaCl_2$ or sulfate for $CaSO_4$. Does the resulting concentration of the anion exceed the MCL for that anion? (see MCLs on the Environmental Protection Agency's web site at: http://www.epa.gov/waterscience/drinking/)

d. If you simply added enough solid $CaCO_3$ to *your* sample to make a saturated solution, would the concentration of Ca^{2+} be high enough to cause CaF_2 to precipitate? (*Hint:* Determine the value of Q and compare it to the K_{sp} of CaF_2. See Exploration 4F for more background information.)

3. Fluoride can also be removed by adsorption to alumina. Between 3 and 5 kg of F^- can be adsorbed per cubic meter (m^3) of alumina. After the alumina is saturated, it can be regenerated by mixing a solution containing hydroxide ions with the solid alumina. For a water supply with fluoride ion concentration of 8 mg/L, how many liters of water can be treated with 1 m^3 of alumina before the alumina must be regenerated?

Laboratory Procedures

The following reagents will be available in the laboratory:

- Acidic alumina
- Neutral alumina
- Basic alumina
- Granular activated carbon
- Powdered activated carbon
- A variety of calcium salts, including 0.005 M solutions of $Ca(NO_3)_2$ and $CaCl_2$, and the solids $CaSO_4$, $Ca(OH)_2$, and $CaCO_3$

Using the general laboratory procedures in **Appendix 6A** and the plan you developed, experiment with these reagents to remove fluoride from your water sample. Consider the following factors when deciding which is the optimum method:

- Cost of the overall system
- Efficiency of the process
- Rate at which the water can be treated
- Simplicity of the overall system

---

note　Because you will be designing your own procedures, it is **essential** that you keep a *detailed, accurate* account of **everything** you do experimentally. If in doubt, write it down! Your final report depends critically on your notes.

Your goal is to optimize each aspect of the treatment plan and come up with a good overall process. Realize that it is impossible to have the perfect system. The general approach requires you to do the following:

- Calculate the expected amount of reagent required to purify your sample, assuming a stoichiometric reaction. You should have done this in the **Developing Ideas** section.
- Use either a batch or flow system to mix 10.0 mL of the water sample with the remediating agent, leave them in contact for a set period of time, and filter the sample to remove any solids (procedures in **Appendix 6A**).

---

note　You may not always see the precipitate form because you are working with small quantities.

- Test the water sample for fluoride concentration using an ion selective electrode (ISE), to determine if the approach has been effective. Procedures can be found in Session 2B on page 28. Don't forget to calibrate the ISE first!
- Adjust the variable you are experimenting with and repeat the experiment.
- Measure the conductivity of the solution to determine the final concentration of total dissolved solids. Does your sample meet the Water Quality Standard for TDS? If not, think about what the contaminant ions might be and how you might reduce their concentration.

Working with the Ideas

After your group has tried a variety of methods to reduce the concentration of fluoride to acceptable levels for drinking water, the group will need to analyze everyone's results to agree on a *preferred* method for removing fluoride. There is no single correct choice. You will need to weigh the cost and effectiveness of each method to make a decision. Be sure you have all of the following information *before you leave the lab*. You will need it to write your report or prepare your poster.

1. What methods did your group test when developing a water treatment plan?
2. What is the preferred method for removing fluoride?
3. Give a step-by-step procedure for the method, in enough detail that someone else could follow it.
4. Describe the chemical reactions occurring during the treatment process by writing the chemical equations and describing your observations.

5. Estimate the cost of your method per million gallons of water, using the table in Appendix 6B on page 139. There are 3.8 liters in a gallon.

6. Why is your preferred method the best of the methods your group investigated?

Exploration 6C

How can excess water hardness be removed from a water supply?

Creating the Context

Why is this an important question?

What do you already know?

Preparing for Inquiry

In many areas of the world, the only available sources of water contain high concentrations of calcium and magnesium (>200 mg/L as $CaCO_3$). In Exploration 2C, you saw that excess hardness (Ca^{2+} and Mg^{2+} ions) in water can cause problems with soap and detergent use and cause scale formation in boilers or other industrial applications. Thus, although water hardness poses no health hazards, communities with high levels of water hardness in their water need to remove it to acceptable levels, usually to the equivalent of less than 100 mg of $CaCO_3$ per liter. The process of removing calcium and magnesium from hard water is referred to as **water softening**. Two methods are used for water softening:

- The lime-soda process
- The ion-exchange process

The lime-soda process relies on precipitation reactions to remove the hardness ions (see Session 5 for background). The ion-exchange process is an adsorption reaction, where the hardness ions adsorb to a surface and thus are removed from the water. In practice, the ion-exchange process is typically used for home water softening, while municipal water treatment plants may utilize either method.

In this Exploration, you will use the plan you developed in Exploration 6A to test different methods for removal of water hardness from your water sample. After experimenting with variables for several laboratory periods, you should be able to answer the Exploration question, *How can excess water hardness be removed from a water supply?*

BACKGROUND READING

Methods for Removing Water Hardness: The Lime-Soda Process

In the lime-soda process, soda ash (Na_2CO_3) and slaked lime ($Ca(OH)_2$) are added to the water to precipitate calcium as $CaCO_3$ and magnesium as $Mg(OH)_2$. Lime and soda ash are very cheap, non-toxic, and readily available.

$$2\,HO^-\,(aq)\ +\ Mg^{2+}\,(aq) \rightleftharpoons Mg(OH)_2\,(s)$$

$$CO_3^{2-}\,(aq)\ +\ Ca^{2+}\,(aq) \rightleftharpoons CaCO_3\,(s)$$

Lime-soda softening is accomplished at municipal water treatment plants by adding lime and soda ash to the water in reactors that have a mixing zone to which the chemicals are added and in which precipitation takes place (see Figure 6-2 on page 114). The water then ?ows into a quiescent zone in which the precipitate is allowed to separate from the softened water by sedimentation. In the laboratory, ?l-tration can also be used to separate the precipitate from the water. The softened water pH is adjusted and the water is ?ltered prior to distribution.

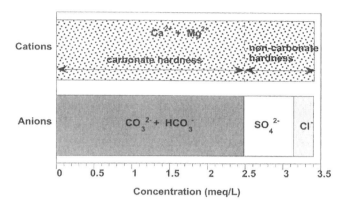

Figure 6-3: The top bar represents the combined concentration of calcium and magnesium in milliequi alents per liter. Carbonate hardness is the amount of calcium and magnesium in the water whose source is the dissolution of minerals such as $CaCO_3$ and $MgCO_3$. Non-carbonate hardness is the amount of calcium and magnesium arising from disso-lution of salts such as $CaSO_4$ and $CaCl_2$.

When using the lime-soda softening process, it is convenient to classify water hardness into **Carbonate Hardness (CH)** and **Non-Carbonate Hardness (NCH)**. **Total Hardness (TH)** is the sum of the two. Carbonate hardness is the portion of the total concentration of Ca^{2+} and Mg^{2+} that arises from dissolution of carbonate min-erals. Non-carbonate hardness is the portion of the total concentration of Ca^{2+} and Mg^{2+} that arises from dissolution of non-carbonate calcium or magnesium salts such as $CaSO_4$ or $MgCl_2$. The bar diagram in Figure 6-3 shows the relationship between carbonate and non-carbonate hardness.

Carbonate hardness is directly related to the total alkalinity of the solution because the carbonate (CO_3^{2-}) and bicarbonate (HCO_3^-) ions are the main sources of alkalinity in natural waters. In most natural waters, the predominant carbonate spe-cies is HCO_3^- because the CO_3^{2-} from dissolved minerals is transformed into HCO_3^- by a reaction with carbonic acid in the water. Carbonic acid is formed when carbon dioxide gas from the air reacts with water.

$$CO_3^{2-}\,(aq)\;+\;H_2CO_3\,(aq)\;\longrightarrow\;2\,HCO_3^-\,(aq)$$

An Example

How do you determine how much lime and soda ash is required for water softening? Consider a typical hard water sample with a total hardness equal to 331 mg of $CaCO_3$ per liter: a CH of 277 mg of $CaCO_3$ per liter plus a NCH of 54 mg of $CaCO_3$ per liter. The pro?le of dissolved constituents is given in Table 6-1, and a bar dia-gram for the sample is shown in Figure 6-4. It is convenient to know concentrations of the constituents in two units: millimoles per liter and milli-equivalents per liter. Using millimoles per liter is helpful when doing calculations involving stoichiomet-ric ratios, and using milliequivalents per liter is helpful when making bar diagrams and understanding the processes. (Milliequivalents are simply the moles of charge per liter; see Exploration 1D for a review of milliequivalents.)

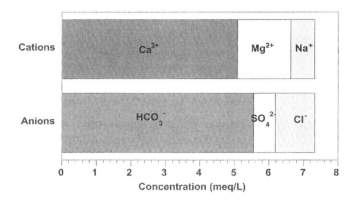

Figure 6-4: Making a bar diagram of a hard water sample is a good wa to determine the dose of lime and soda ash required to soften the water.

Table 6-1   Pro le of a Hard Water Sample

Constituent	Concentration (mol/L)	Concentration (mmol/L)	Concentration (meq/L)
Ca^{2+}	0.00254	2.54	5.08
Mg^{2+}	0.00077	0.77	1.55
Na^+	0.00068	0.68	0.68
HCO_3^-	0.00555	5.55	5.55
Cl^-	0.00114	1.14	1.14
SO_4^{2-}	0.00032	0.32	0.63

Removing magnesium, Mg^{2+}

To carry out a stoichiometric precipitation of Mg^{2+} with hydroxide (OH^-), three reactions need to be considered:

- Lime dissolving in water to produce hydroxide ions.

$$Ca(OH)_2 \text{ (s)} \longrightarrow Ca^{2+} \text{ (aq)} + 2 OH^- \text{ (aq)}$$

- Hydroxide reacting with Mg^{2+} to precipitate solid $Mg(OH)_2$.

$$2 HO^- \text{ (aq)} + Mg^{2+} \text{ (aq)} \longrightarrow Mg(OH)_2 \text{ (s)}$$

- Hydroxide reacting with HCO_3^- to produce CO_3^{2-}.

$$HCO_3^- \text{ (aq)} + HO^- \text{ (aq)} \longrightarrow CO_3^{2-} \text{ (aq)} + H_2O \text{ (l)}$$

What remains is to determine how much lime is required to react completely with both the Mg^{2+} and the HCO_3^- in the sample. Looking at the stoichiometry of the reaction of Mg^{2+} with OH^-, 2 millimoles of hydroxide will need to be added for every millimole of Mg^{2+} in the sample. For the example above, this will be

$$\frac{2 \text{ mmol OH}^-}{\text{mmol Mg}^{2+}} \times \frac{0.77 \text{ mmol Mg}^{2+}}{L} = \frac{1.55 \text{ mm OH}^-}{L}$$

The reaction of hydroxide ions with HCO_3^- to give CO_3^- requires one millimole of OH^- for every millimole of HCO_3^- present in the sample.

$$\frac{1 \text{ mmol OH}^-}{\text{mmol HCO}_3^-} \times \frac{5.55 \text{ mmol HCO}_3^-}{L} = \frac{5.55 \text{ mmol OH}^-}{L}$$

The total amount of hydroxide ion required (in mmol/L) is the sum of these two amounts. For the example above, this value is 7.10 mmol/L. Since $Ca(OH)_2$ (lime) is the source of the hydroxide ion, it provides 2 mmoles of hydroxide per mmole of lime. Thus the "dose" of lime required for a stoichiometric precipitation is:

$$\text{Dose (mmol/L)} = \left(\frac{\text{mmol OH}^-}{L} \text{ for Mg}^{2+} + \frac{\text{mmol OH}^-}{L} \text{ for HCO}_3^- \right) \times \frac{1 \text{ mmol Ca(OH)}_2}{2 \text{ mmol OH}^-}$$

For the example above, the lime dose is 7.10/2 = 3.55 mmol/L.

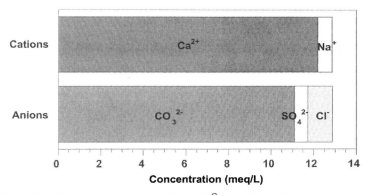

Figure 6-5: After addition of lime, er little Mg^{2+} is left in the solution and the concentration of calcium has increased significantl . All of the bicarbonate has been con erted into carbonate.

After this reaction is complete, a bar diagram of the sample looks like that in Figure 6-5. The amount of calcium has now increased because calcium was added in the form of lime. We must take this fact into account when precipitating the calcium. The milliequivalents of carbonate-containing species have doubled because 5.55 mmol (5.5 meq) of the singly charged species HCO_3^- was converted into 5.55 mmol (11.10 meq) of the doubly charged species CO_3^{2-}.

Removing calcium, Ca^{2+}

The next step is to remove calcium by adding soda ash (Na_2CO_3) to precipitate out calcium carbonate ($CaCO_3$) according to the reaction below:

$$CO_3^{2-} \text{ (aq)} + Ca^{2+} \text{ (aq)} \longrightarrow CaCO_3 \text{(s)}$$

When calculating the soda ash dose required for a stoichiometric precipitation of calcium, you must know the following characteristics of the sample:

• The millimoles of calcium originally in the sample.

• The millimoles of calcium added as lime.

• The millimoles of carbonate originally in the sample as alkalinity.

The soda ash dose is then determined by summing all sources of calcium and subtracting the amount of carbonate already present.

$$\text{Soda ash dose (milllimoles/L)} = [Ca^{2+}]_{orig} + [Ca^{2+}]_{added} - [CO_3^{2-}]_{orig}$$

For the example given above, the soda ash dose is

$$2.54 \text{ mmol/L} + 3.55 \text{ mmol/L} - 5.55 \text{ mmol/L} = 0.54 \text{ mmol/L}$$

The amount of soda ash required per liter is the same as the non-carbonate hardness. In other words, you only need to add soda ash (i.e., total alkalinity) when there is not enough CO_3^{2-} or HCO_3^- in the water to stoichiometrically balance the total hardness. Given the right conditions, such as elevated $[OH^-]$ and/or temperature, the carbonate hardness (i.e., $Ca^{2+} + Mg^{2+}$ = total alkalinity) will "soften itself" without addition of Na_2CO_3. This can be seen inside a kettle used to boil hard water. The white scale is $CaCO_3(s)$ that has formed by the reaction of the Ca^{2+} with HCO_3^- already present in the hard water.

Finally, to determine the ?nal concentration of all ions in the treated water, we need to take into account the stoichiometry of soda ash dissolving in water.

$$Na_2CO_3 \text{ (s)} \longrightarrow 2 Na^+ \text{ (aq)} + CO_3^{2-} \text{ (aq)}$$

The bar diagram for the ?nal treated water is shown in Figure 6-6.

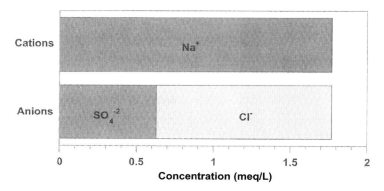

Figure 6-6: Bar diagram of finished water. Compared to the original sample, the concentration of sodium is now higher and the concentration of total dissol ed solids is much lower.

The scenario of reactions just discussed presents the *stoichiometric* picture of lime and soda softening and assumes all reactions are fast *and* that all reactions go to completion. In practice, stoichiometric removal is rarely achieved. Insuf?cient mixing, short reaction times, and the fact that these precipitation reactions do not go to completion at equilibrium prevent complete hardness removal. The minimum hardness that can practically be achieved in municipal water treatment is about 80 mg as $CaCO_3$/L. This level of hardness is perfectly acceptable and, indeed, the complete removal of hardness is not desirable because it results in a very corrosive water.

Using the lime-soda process in municipal water softening yields treated water with a high basicity, i.e., high concentration of $[OH^-]$. Before such water can be ?ltered and distributed, it must be neutralized with acid to bring the water in range of the drinking water standards. Session 8 discusses this aspect of water treatment.

Methods for Removing Water Hardness: The Ion-Exchange Process

In the ion-exchange process, water is allowed to ?ow over an ion-exchange resin (see Figure 6-7). A water softening ion-exchange resin is a polymer with negatively charged sites on the surface that provide locations for positively charged ions such as Na^+, Ca^{2+}, and Mg^{2+} to bind. The resin is initially charged with sodium ions by washing with a solution of sodium chloride (NaCl). As hard water flows over the resin, calcium and magnesium ions bind to the resin while the resin releases sodium ions. When the sodium ions are depleted, the resin can be regenerated by washing with a NaCl solution. Sodium ions are typically used for exchange, since the sodium salts of soap compounds and sodium carbonate are very water soluble. Therefore, water softened by the ion exchange process will not precipitate soap scum on clothes and no scales will form when boiling water. However, sodium can pose a health hazard to people requiring a low-sodium diet. These people should drink bottled water or restrict sodium intake from other sources.

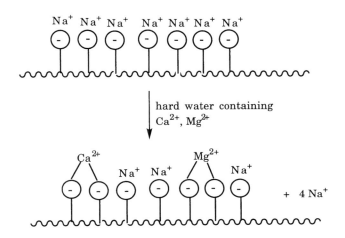

Figure 6-7: Sodium ions are exchanged for calcium and magnesium ions by the ion-exchange process used in home water softening. The spheres in the picture represent a negatively charged group, usually sulfonate, attached to an organic polymer backbone.

Developing Ideas

Unwanted hardness can be removed from water in two ways: the lime-soda process and the ion exchange process. Be sure you have read the **Preparing for Inquiry** section about these two processes before you begin.

One possible way to remove water hardness is by stoichiometric precipitation with lime and soda ash to form the slightly soluble salts $Mg(OH)_2$ and $CaCO_3$ (see Exploration 5C to review the concept of stoichiometric precipitation). The following steps will help you determine how much of the reagents you would need to carry out a stoichiometric precipitation on your sample. Remember that you can remove *more* hardness if you add more reagents (common ion effect, see Exploration 5D), but it will cost more and may not be necessary to remove the required amount of hardness. To begin, you will need to know the following data from the analyses you carried out on your sample in the previous laboratory period:

- The total hardness of your sample in millimoles/L.
- An estimate of the split between Mg^{2+} and Ca^{2+}. A safe bet is that approximately one third of the hardness is from Mg^{2+} on a mole basis.
- The total alkalinity of your sample in mmol/L. For the purpose of these calculations, you may assume that all of the alkalinity exists as bicarbonate, HCO_3^-.

1. Begin by making sure all constituent concentrations (Ca^{2+}, Mg^{2+}, and HCO_3^-) are in millimoles per liter. Fill these values into Column 1 in the following

table. Use the principle of electroneutrality to determine the concentration of "other anions" in solution, assuming a -1 charge on these anions.

Constituent	**Column 1** Concentration in original sample (mmol/L)	**Column 2** Concentration after lime added (mmol/L)	**Column 3** Concentration after soda ash added (mmol/L)
Ca^{2+}			0.00
Mg^{2+}		0.00	0.00
Na^+	0.00	0.00	
HCO_3^-		0.00	0.00
CO_3^{2-}	0.00		0.00
other anions			

a. For a 10 mL sample of your water, calculate the millimoles of lime, $Ca(OH)_2$, required to precipitate the Mg^{2+} as $Mg(OH)_2$, assuming a stoichiometric precipitation. (*Hint:* Write out the equation of the reaction. See pages 122–126 for help.) Determine the millimoles of calcium per liter this step adds to the water.

b. For a 10 mL sample of *your* water, calculate the millimoles of lime, $Ca(OH)_2$, required to convert total alkalinity as bicarbonate (HCO_3^-) to carbonate (CO_3^{2-}). (*Hint:* Write out the equation of the reaction. See pages 122–126 for help.) Determine the millimoles of calcium per liter this adds to the water.

c. For a 10 mL sample of *your* water, calculate the *total* amount of calcium (in millimoles) that will be in the water sample after steps 2 and 3. Don't forget to take into account the calcium originally present in the sample.

d. Fill in the blanks in Column 2 in the table above, assuming steps a and b go to completion.

e. For a 10 mL sample of *your* water, calculate the stoichiometric amount of carbonate required to precipitate all of the Ca^{2+} as $CaCO_3$, assuming complete reaction. (*Hint:* Write out the equation of the reaction. See pages 122–126 for help.)

f. Taking into account the carbonate that was formed in step b, what additional amount of carbonate must be added in the form of soda ash (Na_2CO_3) to remove the calcium?

2. As you learned in Session 4, many reactions reach an equilibrium position where measurable amounts of both products and reactants are present. Precipitation reactions of slightly soluble salts such as $CaCO_3$ and $Mg(OH)_2$ fall in this category; thus, if you add a stoichiometric amount of OH^- intending to precipitate all of the magnesium in your sample, you will not actually remove *all* of the magnesium because the precipitation reaction does not go to completion.

a. Write the solubility product expression for $Mg(OH)_2$. The K_{sp} of $Mg(OH)_2$ is 1.2×10^{-11}. What is the residual concentration of Mg^{2+} in the treated water after stoichiometric precipitation, assuming equilibrium conditions?

b. The K_{sp} of $CaCO_3$ is 8.7×10^{-9}. What is the residual concentration of Ca^{2+} in the treated water after stoichiometric precipitation, assuming equilibrium conditions?

Laboratory Procedures

The following reagents will be available:

- Solid lime, $Ca(OH)_2$
- Solid soda ash, Na_2CO_3
- A saturated solution (0.011 M) of lime
- A 0.05 M solution of soda ash
- Dowex ion-exchange resin, charged with Na^+ ions

Using the general laboratory procedures in Appendix 6A and the plan you developed, experiment with these reagents to remove water hardness from your water sample. Consider the following factors when deciding which method is optimum:

- Cost of the overall system
- Efficiency of the process
- Rate at which the water can be treated
- Simplicity of the overall system

---

 note Because you will be designing your own procedures, it is **essential** that you keep a *detailed, accurate* account of **everything** you do experimentally. If in doubt, write it down! Your final report depends critically on your notes.

Your goal is to optimize each aspect of the treatment plan and come up with a good overall process. Realize that it is impossible to have the perfect system. The general approach requires you to do the following:

- Calculate the expected amount of reagent required to purify your sample, assuming a stoichiometric reaction. You should have done this in the Developing Ideas section.
- Use either a batch or flow system to mix 10.0 mL of the water sample with the remediating agent, leave them in contact for a set period of time, and filter the sample to remove any solids (procedures in Appendix 6A).

---

 note You may not always see the precipitate form because you are working with small quantities.

- Test the water sample for total hardness using an EDTA titration, to determine if the approach has been effective. Procedures can be found in Session 2C on page 34.
- Adjust the variable you are experimenting with and repeat the experiment.
- Measure the conductivity of the solution to determine the final concentration of total dissolved solids. Does your sample meet the Water Quality Standard for TDS? If not, think about what the contaminant ions might be and how you might reduce their concentration.

Working with the Ideas

Problems to be assigned by your instructor

After your group has tried a variety of methods to reduce the water hardness to acceptable levels for drinking water, the group will need to analyze everyone's results to agree on a *preferred* method for removing the water hardness ions Ca^{2+} and Mg^{2+}. There is no single correct choice. You will need to weigh the cost and effectiveness of each method to make a decision. Be sure you have all of the following information *before you leave the lab*. You will need it to write up your report or prepare your poster.

1.   What purification methods were attempted by the members of your group?

2.   What was the preferred method for removing the ion your group was responsible for?

3.   Give a step-by-step procedure for the preferred method, in enough detail that someone else could follow it.

4.   Describe the chemical reactions occurring during the treatment process by writing the chemical equations and describing your observations,

5.   Estimate the cost of your method per million gallons of water, using the table in Appendix 6B on page 139. There are 3.8 liters in a gallon.

6.   Why is your preferred method the best of the methods your group investigated?

Exploration 6D

How can excess iron be removed from a water supply?

Creating the Context

Why is this an important question?

What do you already know?

In Exploration 2D, you saw that excess iron in water is usually a problem with well water because the concentration of soluble iron is much higher in the absence of oxygen. Too much iron results in foul-tasting water and stains on plumbing fixtures. How can communities with high concentrations of iron in their water remove it to levels below the MCL?

In Session 5, you learned about precipitation and adsorption reactions that work to remove soluble contaminants from a water supply. Precipitation reactions require the addition of a precipitating agent and filtration of the resulting precipitate. **Adsorption reactions** require the addition of an insoluble solid with high surface area that the contaminants adhere to, followed by filtration to remove the insoluble solids. Adsorption is the process by which a substance adheres to the surface of an insoluble solid. In general, the greater the surface area of the solid (finer particle size = more surface area), the more sites are available for the substance to adhere to. There are trade-offs, however. The finer the particle size, the slower the water will flow through the system.

Several methods may be used to remove excess iron from water supplies.

- Precipitation and coagulation of ferric hydroxide by adding hydroxide salts
- The ion-exchange process
- Adsorption to activated carbon

In this Exploration, you will use the plan you developed in Exploration 6A to test different methods for removal of iron from your water sample. After experimenting with variables for several laboratory periods, you should be able the answer the Exploration Question *How can excess iron be removed from a water supply?*

Preparing for Inquiry

BACKGROUND READING

Precipitation, Coagulation, and Adsorption

Iron forms very fine, gelatinous precipitates with hydroxide that are nearly impossible to filter. One solution is to incorporate the precipitated iron hydroxide into the structure of another compound that forms a more ordered solid. This process is carried out by adding another insoluble solid, or a reagent that produces an insoluble solid, the **coagulant**. The iron hydroxide particles adhere to the surface of the solid, thus removing the contaminant from the solution.

Common coagulants used in water treatment processes include aluminum sulfate ($Al_2(SO_4)_3$), also called alum, ferric chloride ($FeCl_3$), and polymeric resins with pendant charged groups. These coagulants polarize the suspended particles, which allows them to clump together and form a solid that will settle to the bottom of the treatment container. The combination of precipitation, adsorption, and/or coagulation effectively circumvents unwieldy precipitation reactions.

This method works best for ions that are in low concentration and have very insoluble salts. In some cases, it may be the only way to remove the contaminant quickly without using a very fine filter. However, it requires two remediation agents instead of one, which increases cost and the number of variables that must be optimized. For this reason, fine tuning this procedure may take more time.

Ion-Exchange Process

In the ion-exchange process, water is allowed to flow over an ion-exchange resin (see Figure 6-8). The ion-exchange resin is a polymer with negatively charged sites on the surface that provide locations for positively charged ions such as Na^+, Fe^{3+}, and Ca^{2+} to bind. The resin is initially charged with sodium ions by washing with a solution of sodium chloride (NaCl). As contaminated water flows over the resin, the cations in the water bind to the resin while the resin releases sodium ions. When the sodium ions are depleted, the resin can be regenerated by washing with a NaCl solution. The resins used in water treatment can be regenerated many times.

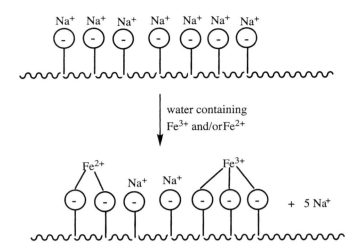

Figure 6-8: Sodium ions are exchanged for iron ions by the ion-exchange process. The spheres in the picture represent a negatively charged group, usually sulfonate, attached to an organic polymer backbone.

Developing Ideas

1. One way to remove iron is by stoichiometric precipitation with hydroxide to form the slightly soluble salt ferric hydroxide, $Fe(OH)_3$ (see Exploration 5C to review the concept of stoichiometric precipitation). The following steps will help you determine which precipitating reagent would be best and how much of the reagent you would need to carry out a stoichiometric precipitation on your sample. Remember that you can remove *more* iron if you add more hydroxide (common ion effect, see Exploration 5D), but it will cost more and may not be necessary to remove the required amount of iron.

    a. Begin by writing the equation for the precipitation reaction of $Fe(OH)_3$. What is the mole ratio of Fe^{3+} to OH^- in the reaction?

    b. Using your results from Session 2D, where you determined the concentration of Fe^{3+} in your sample in mg/L, calculate the concentration of Fe^{3+} in your sample in moles/L.

    c. Calculate the stoichiometric amount of OH^- (in mol) required to remove Fe^{3+} from 10 mL of your sample, assuming that the precipitation reaction goes to completion.

    d. Choose a hydroxide source that will provide the desired amount of OH^- with the least amount of other contaminants added. Check the MCLs on the Environmental Protection Agency's web site at:
    http://www.epa.gov/waterscience/drinking/standards/
    for potentially toxic cations. Available sources of hydroxide for this experiment are
    0.005 M NaOH, 0.005 M KOH, solid $Mg(OH)_2$, and solid $Ca(OH)_2$.

    e. Sodium hydroxide (NaOH) and potassium hydroxide are both available as 0.005 M solutions. Calculate the volume of these solutions (in drops) required to precipitate the Fe^{3+} ion stoichiometrically. There are 20 drops per mL.

    f. The other hydroxide sources ($Mg(OH)_2$ and $Ca(OH)_2$) are available as slightly soluble solids. Look up the K_{sp} values for *one* of these compounds in Appendix 4A and determine the concentration of OH^- in a saturated solution of that compound. Calculate the volume of this saturated solution (in drops) required to precipitate the iron ion stoichiometrically. There are 20 drops per mL.

2. As you learned in Session 4, many reactions reach an equilibrium position where measurable amounts of both products and reactants are present. Precipitation reactions of slightly soluble salts such as $Fe(OH)_3$ fall in this category; thus, if you add a stoichiometric amount of OH^- intending to precipitate all of the iron in your sample, you will not actually remove *all* of the iron.

    a. Write the solubility product expression for $Fe(OH)_3$. The K_{sp} for $Fe(OH)_3$ is 1.1 x 10^{-36}. Will adding a stoichiometric amount of OH^- remove enough iron from your sample so the concentration is below the Maximum Contaminant Level? (See MCLs on the Environmental Protection Agency's web site at http://www.epa.gov/waterscience/drinking/.)

    b. Calculate the residual hydroxide remaining in the solution after stoichiometric precipitation of iron. The acceptable range for hydroxide ion concentrations in drinking water is between $\sim10^{-6}$ and $\sim10^{-8}$ M. Will the residual hydroxide exceed acceptable levels of water hardness?

    c. Calculate the concentration of the counterion that accompanies the hydroxide source you chose to use (e.g., Na^+ if NaOH is used) after the required amount of precipitating reagent is added. Does the resulting concentration exceed the MCL for that ion? (See MCLs on the Environmental Protection Agency's web site at http://www.epa.gov/waterscience/drinking/standards.)

    d. If you simply added enough solid $Mg(OH)_2$ to your sample to make a saturated solution, would the concentration of OH^- be high enough to cause $Fe(OH)_3$ to precipitate? (*Hint:* Determine the value of Q and compare it to the K_{sp} of $Fe(OH)_3$. See **Exploration 4F** for more information.)

3. Iron can also be removed by adsorption to an ion-exchange resin. A total of 9 g of Fe^{3+} can be adsorbed per kg of resin. After the resin is saturated, it can be regenerated by mixing a solution containing sodium ions with the solid resin. For a water supply with an iron concentration of 3 mg/L, how many liters of water can be treated with 1 kg of resin before the resin must be regenerated?

Laboratory Procedures

The following reagents will be available:

- Dowex ion-exchange resin, charged with Na^+ ions
- A variety of hydroxide sources, including 0.005 M NaOH, 0.005 M KOH, solid $Mg(OH)_2$, and solid $Ca(OH)_2$
- The coagulants alum ($Al_2(SO_4)_3$) and ferric chloride ($FeCl_3$)

Using the general laboratory procedures in **Appendix 6A** and the plan you developed, experiment with these reagents to remove iron from your water sample. Consider the following factors when deciding which is the optimum method:

- Cost of the overall system
- Efficiency of the process
- Rate at which the water can be treated
- Simplicity of the overall system

---

note   Because you will be designing your own procedures, it is **essential** that you keep a *detailed, accurate* account of **everything** you do experimentally. If in doubt, write it down! Your final report depends critically on your notes.

Your goal is to optimize each aspect of the treatment plan and come up with a good overall process. Realize that it is impossible to have the perfect system. The general approach requires you to do the following:

- Calculate the expected amount of reagent required to purify your sample, assuming a stoichiometric reaction.
- Use either a batch or flow system to mix 10.0 mL of the water sample with the remediating agent, leave them in contact for a set period of time, and filter the sample to remove any solids (procedures in **Appendix 6A**).

---

note   You may not always see the precipitate form because you are working with small quantities.

- Test the water sample for iron concentration by the colorimetric method, to determine if the approach has been effective. Procedures can be found in Session 2D on page 40.
- Adjust the variable you are experimenting with and repeat the experiment.
- Measure the conductivity of the solution to determine the final concentration of total dissolved solids. Does your sample meet the Water Quality Standard for TDS? If not, think about what the contaminant ions might be and how you might reduce their concentration.

Working with the Ideas

After your group has tried a variety of methods to reduce the concentration of iron to acceptable levels for drinking water, the group will need to analyze everyone's results to agree on a *preferred* method for removing iron. There is no single correct choice. You will need to weigh the cost and effectiveness of each method to make a decision. Be sure you have all of the following information *before you leave the lab*. You will need it to write up your report or prepare your poster.

1. What methods did your group test when developing a treatment plan?

2. What is the preferred method for removing iron from a water supply?

3. Give a step-by-step procedure for the method, in enough detail that someone else could follow it.

4. Describe the chemical reactions occurring during the treatment process by writing the chemical equations and describing your observations.

5. Estimate the cost of your method per million gallons of water, using the table in Appendix 6B on page 139. There are 3.8 liters in a gallon.

6. Why is your preferred method the best of the methods your group investigated?

Making the Link

Looking Back: What have you learned?

Chemical Principles

What chemistry experience have you gained?

As you worked through Session 6, you gained experience in working with some important principles and techniques of chemistry. An understanding of these principles and techniques is valuable in solving a wide range of real-world problems. Your experience with Session 6 should have developed your skills to do the items on the following lists.

Precipitation and Adsorption Reactions

You should be able to:

*   Describe the differences between the batch and flow techniques for water treatment and know the advantages and disadvantages of each (**Exploration 6A**)

*   Determine the stoichiometric dose of a precipitation reagent for removal of fluoride, hardness, or iron (**Explorations 6B, C,** and **D**).

*   Determine the carrying capacity of an adsorbent (**Explorations 6B** and **D**)

*   Recognize that once precipitation has occurred, the final concentration of the precipitated substance in solution is governed by its solubility equilibrium and K_{sp} (**Explorations 6B, C,** and **D**).

*   Recognize that addition of precipitating reagents can introduce additional ions that may be problematic.

Laboratory Measurements

You should be able to:

*   Very accurately use the technique of titration, ion-selective electrode, or colorimetric method to determine the concentration of a contaminant ion.

Thinking Skills

_____

What general skills are you building for your resume?

_____

You have also been developing some general problem-solving and scientific thinking skills that are valued by employers in a wide range of professions and in academia. Here is a list of the skills that you have been building for your resume.

Experimental Design Skills

You should be able to:

- Design a series of experiments to optimize a process.
- Control variables in an experiment.
- Use data to draw conclusions and refine a technique.

Teamwork Skills

You should be able to:

- Work with a team to divide a labor-intensive project in the most efficient way.
- Work with a team to share results and plan the next experiments based on results of prior experiments.

Checking Your Progress

_____

What progress have you made toward answering the Module Question?

_____

Appendix 6A

General procedures for batch and flow system remediation in the laboratory

Making a Filtration Column

The laboratory equivalent of a filtration sand bed in a water treatment plant is a filtration column. This column consists of a 10 mL syringe with one (or more) pieces of filter paper on the bottom. Because all remediation methods either use or generate solids that must be removed, this apparatus is essential for all treatment methods.

1. Clamp a 10 mL syringe in an upright position. Insert the correct diameter filter paper into the syringe by first pushing the paper down (gently!) with a glass rod, then wetting the paper and using the plunger of the syringe to make sure the paper is flush against the sides and bottom of the syringe. Take care in this step; if the filter paper is not flush with the sides, particles can get through.

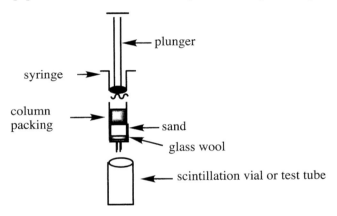

2. Place a layer of glass wool about 1/4-1/2" thick on top of the filter paper and add about 2 mL of sand.

3. Now you have a decision to make. You can use a *flow method* (see p. 113), where a solid remediation agent(s) is weighed out and added directly to the filtration column. A known volume of the contaminated water is allowed to flow through the syringe column. Or you can use a *batch method* (see p. 114), where a known volume of the contaminated water and a known amount of remediation agent (solid or liquid) are combined in a small beaker and stirred for an appropriate amount of time. The syringe column is used to filter the mixture, separating water from the remediation agent.

4. Once you have decided which method to use, follow the procedures for the chosen method, given next. Every time you do a test, make note of the following:

    *   the moles of remediation agent used, including the mass of reagent if it is a solid and the volume and concentration of reagent if it is a solution
    *   the volume of water treated
    *   the flow rate through the system, if it is a flow remediation
    *   the contact time

A Caveat

Some solid remediation agents and/or precipitates may pass through the sand, glass wool, and filter paper. If so, experiment with other filtration set-ups, or increase the amount of sand and/or glass wool in your column. As a last resort, use the *very expensive* syringe filters. Remember, the final product should not contain any particulate matter.

FLOW SYSTEM REMEDIATION

1. Prepare a syringe filtration column as described above

2. Add a known mass of solid remediating reagent on top of the sand layer. For reagents that work by adsorption (charcoal, alumina), start with 0.1-0.5 g of reagent for 10 mL of water. For reagents that work by precipitation, use the stoichiometric amount of reagent you calculated in your prelab assignment. You may wish to start with the stoichiometric amount and then add an excess of reagent if the stoichiometric amount doesn't work.

3. Add another 0.5 mL of sand to the top of the column to keep the bed of remediation agent from being disturbed by pouring water on it.

4. Now add a known volume (approximately 10 mL) of your sample to the prepared column, taking care not to add too much sample at once.

5. Collect the sample in a clean, dry container, measuring the volume of the remediated sample and measuring the flow rate to determine the contact time. You will need about 8 mL of remediated sample. Label the treated water and note the conditions used for treatment in your laboratory notebook, then proceed to the instructions for determination of your contaminant ion.

   - Fluoride analysis: Session 2B, page 28.
   - Hardness analysis: Session 2C, page 34
   - Iron analysis: Session 2D, page 40

BATCH REMEDIATION

1. Prepare a syringe filtration column as described above.

2. In a beaker, mix a known amount of remediation agent and a known volume of contaminated water. For reagents that work by adsorption (charcoal, alumina), start with 0.1-0.5 g of reagent per 20 mL of water. For reagents that work by precipitation, use the stoichiometric amount of reagent you calculated in your prelab assignment. You may wish to start with the stoichiometric amount and then try adding an excess of reagent if the stoichiometric amount doesn't work.

- Swirl the mixture around in the beaker or stir it with a clean stirring rod and allow most of the particles to settle. Try to avoid putting particulates onto the filtration column to prevent clogging of the column.

---

 note   You may not always see the precipitate form because you are working with small quantities of reagents.

---

3. Note the contact time, then decant the solution into the filtration column.

4. Collect the sample in a clean, dry container, measuring the volume of the remediated sample and measuring the flow rate to determine the contact time. You will need about 8 mL of remediated sample for analysis. Label the treated water and note the conditions used for treatment in your laboratory notebook, then

proceed to the instructions for determination of your contaminant ion to see if your method was successful.

- **Fluoride analysis:** Session 2B, page 28.
- **Hardness analysis:** Session 2C, page 34
- **Iron analysis:** Session 2D, page 40

Appendix 6B

Cost of Water Treatment Reagents

Chemical	Price per kg (unless otherwise noted)
Activated carbon, granular	$2
Activated carbon, powdered	$2
Alum, $Al_2(SO_4)_3$	$0.18
Alumina, acidic	$2.50
Alumina, basic	$2.50
Alumina, neutral	$2.50
Calcium chloride, $CaCl_2$	$2
Calcium carbonate, $CaCO_3$	$10
Calcium sulfate, $CaSO_4$	$1.80
Ferric chloride, $FeCl_3$	$0.25
Ferrous sulfate, $Fe_2(SO_4)_3$	$0.25
Hydrochloric acid, HCl (12 M soln.)	$1.00 per liter
Ion-exchange resin	$3
Lime, $Ca(OH)_2$	$0.50
Magnesium hydroxide, $Mg(OH)_2$	$1.50
Soda ash, Na_2CO_3	$1.20
Sodium hydroxide, NaOH	$1.80

Acids and Bases I

What are acids and bases?

Exploration 7A: The Storyline

Creating the Context

Why is this an important question?

When you initially laid out a plan for purifying a water supply, you probably mentioned the need for disinfection of the water and removal of toxic substances like metal ions. Acids and bases are used extensively in these water treatment processes. The goals of this Session are to acquaint you with the structures of acids and bases, to learn about the connection between structure and the equilibrium position of acid-base reactions, and to learn how to use equilibrium expressions to determine or predict concentrations of acids and bases in aqueous solutions. By the end of this Session, you should be able to answer the question *What are acids and bases?*

We encounter many compounds in our daily lives that are classified as acids or bases. For example, vinegar is a dilute solution of acetic acid, citrus fruit juices contain citric acid, and the hydrochloric acid produced in the stomach helps to digest food. Natural rainwater can be acidic due to the presence of acids produced by the dissolution of the atmospheric gases CO_2, SO_2, and NO_2. Common household bases include soaps, ammonia, lye, bleach, and baking soda.

What do you already know?

You are probably familiar with some properties of the acids and bases. Acids (like citric acid in lime juice) taste sour or tart, whereas bases (like soap) feel slippery and can taste bitter. There are also notable *chemical* similarities among acids and among bases. In this Exploration, you will examine some compounds and make observations about the characteristics that define a compound as acidic or basic.

Developing Ideas

Your instructor will add "universal indicator" to 0.01 M solutions of the compounds in Table 7-1. The indicator allows you to determine the acidity or basicity of the solutions by observing the color of the solution. A light bulb conductivity meter will be used to measure whether the solution contains ions. Your goal is to make observations, look for trends, and generalize about compounds that are acidic or basic.

Table 7-1   Acidic, Basic, and Neutral Compounds

Key to Indicator Color

Strongly acidic	pink
Weakly acidic	orange
Neutral	yellow
Weakly basic	green
Srongly basic	blue

Key to Ionization

Bright (++)	extensive
Dim (+)	partial
No light (-)	minimal

Compound	Structure	Ions formed?	Acid, neutral, or base?
CH_3OH	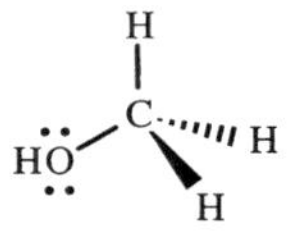		
CH_3COCH_3	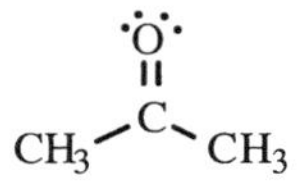		
H_2CO_3	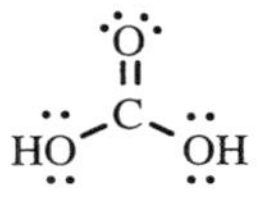		
CH_3COOH	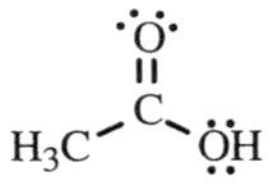		
Na (s)			
HCl	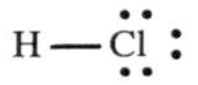		
$N(CH_3)_3$	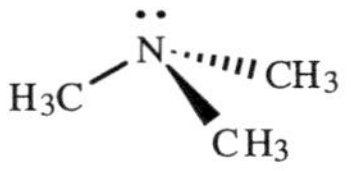		
$Ca(OH)_2$	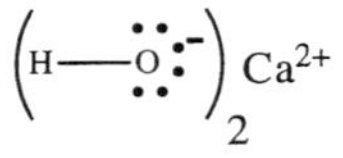		
H_2SO_4	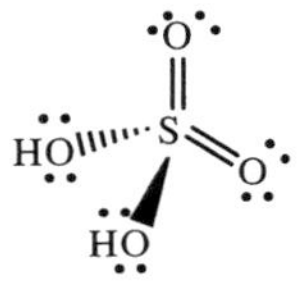		
NaCl	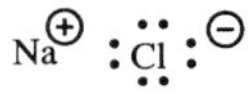		
NH_3	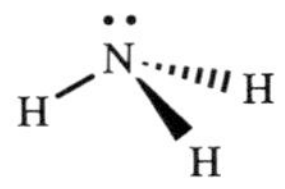		
$NaHCO_3$	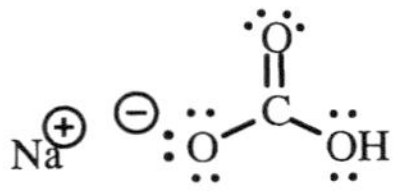		
NaOH	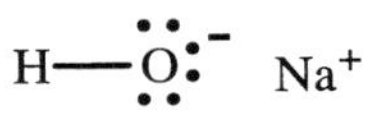		

1.   Group the compounds in two ways:

    a.   By the color of the indicator. Use the chart at the left of the table to classify the solutions as strongly acidic, weakly acidic, neutral, weakly basic, or strongly basic.

    b.   By the amount of ionization, as indicated by the amount of light produced by the light bulb.

2.   Look at the structures of the compounds. What structural features do the acids have in common? the bases?

3.   Why are ions produced in some reactions and not in others? What guesses can you make about this?

Working with the Ideas

The following problems will help you understand the concepts in more depth.

4.   Based on your work in the **Developing Ideas** section, classify the following compounds as acids, bases, or neutral compounds. Begin by drawing the Lewis dot structure and the shape of each molecule.

    a. H_3PO_4 (H's bound to O's)

    b.   KBr

    c.   HOCl

    d. PH_3

    e.   RbOH

    f. H_2O

    g.   HOOC-COOH

    h. $NaNO_3$

Looking Ahead

Several Exploration Questions can guide you to explore the relationship of acids and bases to water treatment. These will be used at the discretion of your instructor.

•   **Exploration 7B:**   How are acid and base strengths correlated to the extent of the acid-base reaction?

•   **Exploration 7C:**   How can we best quantify acid and base concentrations?

•   **Exploration 7D:**   How do you determine equilibrium concentrations of strong and weak acids and bases?

•   **Exploration 7E:**   How is acid strength related to the structure and composition of the acid?

•   **Exploration 7F:**   What are the energetics of acid-base reactions?

Exploration 7B

How are acid and base strengths correlated to the extent of the acid-base reaction?

Creating the Context

Why is this an important question?

As you saw in Exploration 7A, acids and bases are not all the same strength. Different water treatment applications can make use of both strong and weak acids and bases. For example, hydrochloric acid (HCl), sometimes called muriatic acid, a strong acid, is used to clean tile surfaces. A weaker acid such as carbonic acid (H_2CO_3) would not make as powerful a cleaner; however, carbonic acid can serve other purposes. This compound, in conjunction with the bicarbonate and carbonate ions (HCO_3^- and CO_3^{2-}) is nature's choice for regulating the acidity of natural waters, something a strong acid cannot do. You will need to know more about acid and base strength to understand where certain acids and bases will be most useful in water treatment processes.

In your work in Exploration 7A, you observed that compounds that did not light up the light bulb when they dissolved in water were not acidic or basic, but produced neutral solutions. This should give you a hint as to one criterion for the definition of acidity or basicity in aqueous solution. *In order for a compound to be acidic or basic in aqueous solution, the compound must either dissociate in water or react with water to form ions.* However, not all compounds that form ions will make an aqueous solution acidic or basic. For example, NaCl forms Na^+ and Cl^- ions when it dissolves, yet the solution is neutral. A more specific definition is that *acidic compounds form hydronium ions, H_3O^+ when dissolved in water, and basic compounds form hydroxide ions, OH^-, when dissolved in water.* In this Exploration, we will examine these criteria for acids and bases in detail, with particular attention to the extent of the reaction that forms ionic species. By the end of the Exploration, you should be able to answer the question, *How are acid and base strengths correlated to the extent of the acid-base reaction?*

Preparing for Inquiry

BACKGROUND READING

Acids and bases are commonly defined using the Brønsted-Lowry convention, proposed in 1932 by Johannes Brønsted (1879-1947) and Thomas Lowry (1879-1936). This definition works well for acids and bases in aqueous solutions such as those we are studying. The Brønsted-Lowry theory states that an acid is a substance that can donate a proton (H^+) to the solvent (water), and that a base is a substance that can accept a proton from water. Thus an **acid** is defined as a **proton donor** and a **base** is defined as a **proton acceptor**.

Generally speaking, we can define an acid-base reaction for the general acid, HA according to the reaction written below. In the reaction, water is acting as a base, accepting a proton. The products of the reaction are the **conjugate acid**, H_3O^+, and the **conjugate base**, A^-.

$$HA + H_2O \underset{}{\overset{K_a}{\rightleftharpoons}} H_3O^+ + A^-$$

$$\text{acid} \qquad \text{base} \qquad \begin{array}{c}\text{conjugate}\\\text{acid}\end{array} \qquad \begin{array}{c}\text{conjugate}\\\text{base}\end{array}$$

An acid-base reaction for the general base, B, can be defined according to the reaction below.

$$B: \quad + \quad H_2O \underset{}{\overset{K_b}{\rightleftharpoons}} BH^+ \quad + \quad OH^-$$

base   acid   conjugate   conjugate
acid   base

In this case, water is acting as an *acid*, donating a proton to the base. The products of the reaction are the conjugate acid, BH^+, and the conjugate base, OH^-.

Both of these reactions are written as equilibria, with an associated equilibrium constant. The convention is to use K_a to designate **acid ionization constants** and K_b to designate **base protonation constants**. Tables of K_a and K_b values for selected compounds can be found in Appendix 7A. As with solubility equilibria, these equilibrium constants provide information about the *extent* to which the reactions proceed to products.

For a compound to be an acid in the Brønsted-Lowry sense, it must have a hydrogen atom bound to an electronegative atom, X, where X = O, N, F, Cl, Br, I, or S. The polarization of the H-X bond caused by the difference in electronegativity between H and X makes it easy for an electron-rich substance (a base) to pluck off the proton. See Table 7-3 and Table 7-4 in Appendix 7A for examples.

$$\overset{\longmapsto}{\underset{\delta^+ \quad \delta^-}{H-X}}$$

A Brønsted-Lowry base must have an unshared pair of electrons available for bonding to a proton. Often this unshared pair resides at a negatively charged site on the molecule, usually on an electronegative atom such as oxygen or nitrogen. The hydroxide ion, OH^-, is an example of this type of base. The unshared pair of electrons can also reside on a neutral atom, such as the nitrogen atom of ammonia (NH_3).

The species in aqueous solution that is responsible for basicity is the hydroxide ion, OH^-. Strong bases are either salts that contain the hydroxide ion or compounds that react with water to form hydroxide ions. Hydroxides of the Group 1 alkali metals Li, Na, K, Rb, and Cs, and the Group 2 alkaline earth metals Mg, Ca, and Sr are all strong bases. The Group 1 hydroxides produce one mole of hydroxide ions per mole of the substance dissolved in water.

$$KOH\ (s) \longrightarrow K^+\ (aq) \ + \ OH^-\ (aq)$$

For the group 2 hydroxides, two moles of hydroxide ions are produced per mole of the substance dissolved in water.

$$Mg(OH)_2\ (s) \rightleftharpoons Mg^{2+}\ (aq) \ + \ 2\ OH^-\ (aq)$$

Even so, the basicity of solutions of these compounds is limited by the insolubility of these compounds in water, as dictated by their K_{sp} values (see Appendix 4B). In contrast, the hydroxides of the Group 1 metals dissociate essentially completely in water, resulting in higher concentrations of hydroxide.

What background information is helpful?

The hydroxide ion can also be produced by reaction of a basic compound with water. For example, ammonia reacts with water to remove a proton and produce NH_4^+ and OH^-.

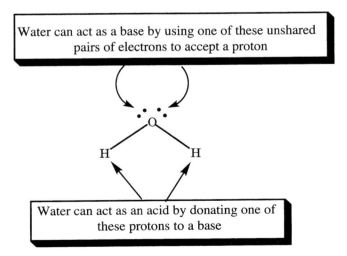

Water as an Acid and a Base

Because we are studying acid-base reactions in *aqueous* solution, water is the solvent, and the most readily available reactant. Thus all acid-base reactions in aqueous solution are defined in terms of reactions with the water molecule.

You may have noticed that water is unusual—it can act both as an acid and as a base. Water reacts with acids to *accept* a proton, using one of its unshared pairs of electrons. Water can also react with bases to *donate* a proton (see Figure 7-1). In fact, water can act as a proton donor and proton acceptor towards itself.

The equilibrium expression for this reaction is written

$$K_w = [H_3O^+] [OH^-] = 1.0 \times 10^{-14} \text{ at } 25°C$$

Because the concentration of H_2O in dilute aqueous solutions is large, we can assume that it remains essentially constant; thus H_2O does not appear in the equilibrium expression. The equilibrium constant K_w is called the **ion-product constant** of water. When $[H_3O^+] = [OH^-]$, the solution is *neutral*. When $[H_3O^+]$ is larger than $[OH^-]$, the solution is *acidic*, and when $[OH^-]$ is larger than $[H_3O^+]$, the solution is *basic*.

Water can act as a base by using one of these unshared pairs of electrons to accept a proton

Water can act as an acid by donating one of these protons to a base

Figure 7-1: *The water molecule can act as both an acid and a base.*

How can you find out more?

To learn more, we will examine a collection of acids and bases, looking for trends in acidity as they relate to the extent of the reaction with water.

Developing Ideas

In this activity, you will examine a variety of conjugate acid/base pairs and make observations that will lead you to some generalizations about acid and base strength and extent of reaction.

1. For each of the compounds in the table below, observe the demonstration by your instructor and do the following:

    a. Classify the compounds as strong or weak acids or bases, as indicated by the color of the indicator

    b. For the acids and bases (not the salts), note the extent of the reaction, as indicated by the intensity of the light bulb.

Key to Indicator Color

Strongly acidic	pink
Weakly acidic	orange
Neutral	yellow
Weakly basic	green

Key to Ionization

Bright (++)	extensive
Dim (+)	partial
No light (-)	minimal

Compound		Strong or weak acid? Strong or weak base?	Extent of reaction (as indicated by ionization)
CH_3OH	acid		
$Na^+ CH_3O^-$	conjugate base		---------
CH_3COOH	acid		
$Na^+ CH_3COO^-$	conjugate base		---------
HCl	acid		
$Na^+ Cl^-$	conjugate base		---------
NH_3	base		
$NH_4^+ Cl^-$	conjugate acid		---------
$Na^+ OH^-$	base		---------
H_2O	conjugate acid		
H_2CO_3	acid		
$Na^+ HCO_3^-$	conjugate base, #1		---------
$(Na^+)_2CO_3^{2-}$	conjugate base, #2		---------
H_2SO_4	acid		
$Na^+ HSO_4^-$	conjugate base, #1		---------
$(Na^+)_2 SO_4^{2-}$	conjugate base, #2		---------

2. How does the extent of ionization relate to the strength of an acid or base?

    • More ionized means the acid or base is (stronger/weaker)

    Explain your reasoning.

3. Write the equilibrium expression for the generic reaction of an acid with water. What generalization can you make about the relationship between the extent of the ionization reaction and the numerical value of K_a?

    • More ionized means the K_a is (larger/smaller)

    Explain your reasoning.

4. What other generalizations can you make about conjugate acid/base pairs?

    a. If an acid (HA) is strong, the conjugate base (A⁻) is a (strong/weak) base.

    b. If a base (B) is strong, the conjugate acid (BH⁺) is a (strong/weak) acid.

    c. If an acid (HA) is weak, the conjugate base (A⁻) is a (strong/weak) base.

    d. If a base (B) is weak, the conjugate acid (BH⁺) is a (strong/weak) acid.

5. The value of K_w is 1.0×10^{-14} at 25°C.

   a. Is water a strong acid or a weak acid?

   b. Is water a strong base or a weak base?

   c. What is the concentration of H_3O^+ in pure water at 25°C?

Working with the Ideas

The following problems will help you understand the concepts in more depth.

6. Compare carbonic acid (H_2CO_3) and acetic acid (CH_3COOH).

   a. Write balanced chemical equations for the reactions of these compounds with water. Use structurally accurate representations of the compounds, including the unshared electron pairs.

   b. Write the equilibrium expressions for the reactions of these acids with water.

   c. Look up the values of K_a in Table 7-4 in Appendix 7A. Which is the stronger acid? Explain how you know.

7. Compare the two bases ammonia (NH_3) and methyl amine (NH_2CH_3).

   a. Write balanced chemical equations for the reactions of these compounds with water. Use structurally accurate representations of the compounds, including the unshared electron pairs.

   b. Write the equilibrium expressions for the reactions of these bases with water.

   c. Look up the values of K_b in Table 7-5 in Appendix 7A. Which is the stronger base? Explain how you know.

8. The bicarbonate ion, HCO_3^-, can act as either an acid or a base.

   a. Write the chemical equation for HCO_3^- acting as an acid.

   b. Write the chemical equation for HCO_3^- acting as a base.

   c. Look up the values of K_b for HCO_3^- and CO_3^{2-} in Table 7-5 in Appendix 7A. Which compound is the stronger base? Explain how you know..

   d. Which is the stronger acid, H_2CO_3 or HCO_3^-?

9. A solution contains H_3O^+ at a concentration of 0.0001 M. What is the concentration of OH^-?

Exploration 7C

How can we best quantify acid and base concentrations?

Creating the Context

Why is this an important question?

In Exploration 7A, you examined a series of compounds and looked for trends in their compositions that correlated with acidity or basicity, labeling the compounds as weakly acidic or basic, strongly acidic or basic, or neutral. The color of an acid-base indicator guided your choices. While such generalizations are useful, we often need to know *exactly* how much H_3O^+ or OH^- is in solution. In this Exploration, you will examine two methods for representing acid and base concentrations and gain some experience using them. By the end of the Exploration, you will be able to answer the question *How can we best quantify acid and base concentrations?*

So far, you have reported concentrations of species in solution in moles per liter or mg/L. These units provide a direct measurement of concentrations and help when comparing concentrations of approximately the same magnitude. For example, it's fairly easy to understand the comparison between a water supply that contains 325 mg/L of hardness and one that contains 72 mg/L of hardness. But what are the implications of concentrations differing by orders of magnitude (powers of 10), such as an acid that is 1×10^{-6} M compared to another that is 1×10^{-4} M?

Another way to report concentrations is to take the negative logarithm of the concentration, symbolized by the letter "p". This method is used to quantify the concentration of acid or base in solution, where the **pH** represents the negative logarithm (base 10) of the H_3O^+ concentration in moles per liter.

$$pH = -\log_{10}[H_3O^+]$$

Recall that $K_w = [H_3O^+][OH^-] = 1.0 \times 10^{-14}$ at 25°C. Taking the negative base 10 logarithm of each of the terms in this expression and using the notation pX to signify $-\log_{10}[X]$, we can use the laws of logarithms to write

$$pK_w = pH + pOH = 14$$

where

$$pH = -\log[H_3O^+]$$

$$pOH = -\log[OH^-]$$

$$pK_w = -\log K_w$$

When $[H_3O^+] = [OH^-]$, the solution is neutral and pH = pOH = 7. Converting this number to concentration units gives $[H_3O^+] = [OH^-] = 1.0 \times 10^{-7}$ M in pure water at 25°C.

Because acidic solutions contain a higher concentration of hydronium ions than hydroxide ions, the pH of an acidic solution will be less than 7 at 25°C. For example, if the concentration of H_3O^+ is 1.0×10^{-6} M, the pH is

$$-\log(1.0 \times 10^{-6}) = 6.0$$

Because basic solutions contain a lower concentration of hydronium ions than hydroxide ions, the pH of a basic solution will be greater than 7. For example, if the concentration of H_3O^+ is 1.0×10^{-8} M, the pH is

$$-\log(1.0 \times 10^{-8}) = 8.0$$

Table 7-2 shows the pH of some common solutions you may have encountered in your everyday life.

Taking the negative logarithm of a number is also used when expressing equilibrium constants. For example, the pK_a of an acid is the negative logarithm of its acid dissociation constant. For acetic acid, with a K_a of 1.74×10^{-5}, the pK_a is 4.76. The net effect of taking the negative log of the number is to take a very small (or very large) number and transform it into a single- or double-digit number.

In this Exploration, we'll compare the two systems of reporting acid concentration (moles/L vs. pH) by looking at situations in which one system might be more advantageous than the other. We'll also examine the relationship between $[OH^-]$ and $[H_3O^+]$ in more detail.

What background information is helpful?

How can you find out more?

Table 7-2   pH of Common Solutions

Solution	Acid or Base	pH
Battery acid	Sulfuric acid	0.3
Gastric juice	Hydrochloric acid	1.0–2.0
Lemon juice	Citric acid	2.4
Commercial vinegar	Acetic acid	3.0
Orange juice	Citric acid	3.5
Urine	Uric acid	4.8–7.5
Rainwater	Carbonic acid	5.5-6.0
Milk	Lactic acid	6.5
Saliva	Bicarbonate and proteins	6.4–6.9
Pure water	---	7.0
Blood	Bicarbonate ion	7.35–7.45
Baking soda solution	Bicarbonate ion	8.5
Household ammonia	Ammonia	11.5
Household lye	Sodium hydroxide	13.6

Developing Ideas

1. The following data show how the concentration of H_3O^+ varies with a change in $[OH^-]$.

$[OH^-]$	$[H_3O^+]$	$-\log [OH^-]$	$-\log [H_3O^+]$
1	1×10^{-14}		
1×10^{-2}	1×10^{-12}		
1×10^{-4}	1×10^{-10}		
1×10^{-6}	1×10^{-8}		
1×10^{-8}	1×10^{-6}		
1×10^{-10}	1×10^{-4}		
1×10^{-12}	1×10^{-2}		
1×10^{-14}	1		

a. On graph paper, plot $[OH^-]$ vs. $[H_3O]^+$

b. Calculate $-\log [OH^-]$ and $-\log [H_3O]^+$ and add these values to the table.

c. Plot $-\log [OH^-]$ vs. $-\log [H_3O]^+$

d. Compare your plots. What are the advantages of using a log scale for pH?

2. Take the negative logarithm of the following molar concentrations of Ca^{2+} in a water supply. The concentrations in mg/L are given for reference only.

Concentration of $CaCO_3$ (mg/L)	Concentration of Ca^{2+} (mol/L)	$-log[Ca^{2+}]$
75	0.00075	
210	0.00210	
135	0.00135	
350	0.00350	
500	0.00500	

You have just calculated Ca^{2+} concentrations over the range of typical values found in natural waters. What might be a limitation to using the log scale for this application?

3. Look in Appendix 7A at the acid ionization constants for weak acids and strong acids.

    a. What is the relationship between pK_a and acid strength?

    b. What is the advantage of using pK_a instead of K_a to report acid strengths?

    c. What might be a disadvantage of using pK_a instead of K_a to report acid strengths?

Working with the Ideas

The following problems will help you understand the concepts in more depth.

4. In any aqueous solution, the concentrations of hydronium ions and hydroxide ions are related to each other by the ion-product constant, K_w. The plot in Figure 7-2 describes the relationship between $[H_3O^+]$ and $[OH^-]$ over a small range of concentrations near neutral.

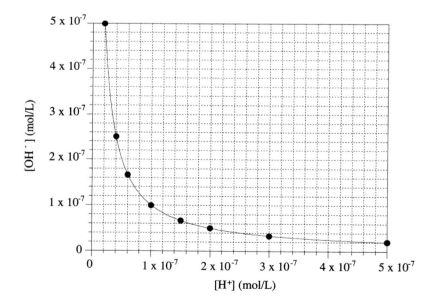

Figure 7-2: In dilute aqueous solution, the relationship between the concentration of H_3O^+ and OH^- at equilibrium is defined by the equilibrium expression. Any point on the curve defines the system under equilibrium conditions.

  a. Write the equilibrium expression for the ionization of water.

  b. At what point on the curve is the solution neutral? Define neutral.

  c. At what points on the curve is the solution acidic? Define acidic.

  d. At what points on the curve is the solution basic? Define basic.

5. Use the plot to *estimate* the answers to the following questions.

  a. When the concentration of OH^- is 2×10^{-7} M, what is the concentration of H_3O^+? Is the solution acidic, basic, or neutral?

  b. When the concentration of H_3O^+ is 3.5×10^{-7}, what is the concentration of OH^-? Is the solution acidic, basic, or neutral?

6. Use the equilibrium expression for the ionization of water to calculate the *exact* concentrations of $[H_3O^+]$ and $[OH^-]$ for the questions in problem 5.

7. Calculate the pH and the pOH of the following solutions:

  a. 1.2 M H_3O^+

  b. 6.4×10^{-8} M H_3O^+

  c. 1.5×10^{-3} M OH^-

  d. 2×10^{-12} M OH^-

  e. 462 mg/L OH^-

  f. 3 g/L OH^-

Exploration 7D

How do you determine equilibrium concentrations of strong and weak acids and bases?

Creating the Context

Why is this an important question?

Because strong and weak acids act differently in water treatment reactions, we need to be able to predict equilibrium concentrations of all species involved in an acid-base reaction. This process requires using your knowledge of equilibrium expressions, some algebraic calculations, and an ability to predict whether an acid or base is strong or weak. This Exploration will give you some practice doing these calculations and show you some shortcuts. By the end of the Exploration, you should be able to answer the question *How do you determine equilibrium concentrations of strong and weak acids and bases?*

Your work in Exploration 7B allowed you to generalize that the reaction of *strong acids* with water goes nearly to completion to form the hydronium ion, H_3O^+ and the conjugate base of the acid. As a result, strong acids have large K_a values (greater than 1) and small (even negative) pK_a values (see Table 7-3 in Appendix 7A). Thus, the moles of H_3O^+ formed and the moles of A^- formed from a strong

What do you already know?

acid, HA, are both essentially equal to the moles of the acid HA_s initially added to the solution.

$$HA_s \ + \ H_2O \longrightarrow H_3O^+ \ + \ A_s^-$$

strong
acid

Recall that *weak acids* do *not* dissociate or ionize completely, leaving significant amounts of undissociated acid, HA_w, in solution.

$$HA_w \ + \ H_2O \rightleftharpoons H_3O^+ \ + \ A_w^-$$

weak
acid

For weak acids, K_a values are small (less than 1) and pK_a values are large (see Table 7-4 in Appendix 7A). This situation is very similar to the dissolution of slightly soluble salts in that the reaction does not go much toward completion. Instead, the reactants are favored and a mixture of products and reactants is present at equilibrium. The equilibrium expression provides a means for calculating exact concentrations of the species remaining in solution.

$$K_a = \frac{\left[H_3O^+\right]\left[A^-\right]}{\left[HA\right]}$$

Recall from Exploration 7B that *strong bases* are typically salts like NaOH or KOH that dissociate essentially completely in water to form hydroxide ions.

$$KOH \ (s) \longrightarrow K^+ \ (aq) \ + \ OH^- \ (aq)$$

Weak bases react only partially with water, resulting in a mixture of reactants and products at equilibrium.

$$B_w \ + \ H_2O \ \rightleftharpoons B_wH^+ \ + \ OH^-$$

weak
base

The equilibrium expression provides a means for calculating exact concentrations of the species remaining in solution.

$$K_b = \frac{\left[BH^+\right]\left[OH^-\right]}{\left[B\right]}$$

In general, K_b values are small for weak bases, typically less than 1.

To learn more about determining exact concentrations of these species in solution, we will examine some acid-base reactions.

Developing Ideas

1. Muriatic acid, HCl, is widely used in water treatment plants, both as a cleaning agent and as a reagent to neutralize bases. Your mission in this problem is to determine the pH of a 0.25 M solution of HCl.

   a. Begin by determining if HCl is a strong or weak acid. You may need to use the tables in Appendix 7A for more information. Reread the **Creating the Context** section if you are unsure of where to begin.

   b. Write the equation for the reaction of HCl with water. What is the stoichiometric relationship between moles of HCl dissociated and moles of H_3O^+ produced?

c. Use the information from **a** and **b** to fill in the table below.

	HCl(aq) (mol/L)	H_3O^+ (aq) (mol/L)	Cl^- (aq) (mol/L)
Initial concentration. (before any dissociation)			
Final equilibrium con- centration			

d. Write the equilibrium expression for the reaction of HCl with water., and use the data table to substitute concentrations into the equilibrium expression and solve for $[H_3O^+]$.

e. Calculate the pH of the solution.

2. Hypochlorous acid, HOCl, is produced in water when disinfecting agents such as Cl_2 and NaOCl are added. Your mission in this problem is to determine the pH of a 0.25 M solution of HOCl.

a. Begin by determining if HOCl is a strong or weak acid. You may need to use the tables in Appendix 7A for more information. Reread the **Creating the Context** section if you are unsure of where to begin.

b. Write the equation for the reaction of HOCl with water. What is the stoichiometric relationship between moles of HOCl dissociated and moles of H_3O^+ produced?

c. You may not know yet exactly how many H_3O^+ and OCl^- ions actually dissolved in the solution, but we can represent this unknown amount with an *x*. Because you know the relationship between the moles of HOCl dissociated and the moles of H_3O^+ and OCl^- ions produced, you now have a system with just one unknown. Use this information to fill in the table below.

	HOCl(aq) (mol/L)	H_3O^+ (aq) (mol/L)	OCl^- (aq) (mol/L)
Initial concentration. (before any dissociation)			
Final equilibrium con- centration (in terms of *x*)			

d. Write the equilibrium expression for the reaction of HOCl with water, and use the data table to substitute concentrations into the equilibrium expression.

e. You are now confronted with a situation in which making an approximation might save you some work. Compare the magnitude of the K_a of HOCl with the concentration of HOCl. Are they different by more than three orders of magnitude (powers of 10)?

f. How significantly would the denominator of your equilibrium expression be altered by ignoring *x*. (i.e., assuming *x* is small relative to 0.25)? Solve the problem for *x* to answer the question.

g. Calculate the pH of the solution. How does it compare to the pH of the 0.25 M HCl solution in problem 1? Explain.

Working with the Ideas

The following problems will provide you with some practice with the new concepts.

3. Acetic acid (CH_3COOH) dissociates in water according to the reaction

CH₃COOH (aq)	H₂O (l)	CH₃COO⁻ (aq)	H₃O⁺ (aq)

The equilibrium constant, K_a, for the reaction is 1.75×10^{-5}. If you add 0.2 moles of acetic acid to 1 L of water, what will be the equilibrium concentrations of CH_3COOH, CH_3COO^-, and H_3O^+?

4. How much 12 M HCl must be added to 1 liter of pure water to obtain a cleaning solution with a concentration of $[H_3O^+]$ equal to 0.1 M?

5. A 0.1 M solution of formic acid (HCOOH) is prepared. What is the pH of this solution?

Exploration 7E

How is acid strength related to the structure and composition of the acid?

Creating the Context

Why is this an important question?

When choosing an acid for a water treatment application, we need to know whether it is a strong or weak acid. Many analytical procedures for water quality analysis require the presence of an acid that does not have a reactive (basic) counterion. Other applications require a basic counterion to serve as a buffer (Session 9) to maintain a constant pH during the analysis. Hence, water treatment engineers need to be able to select an appropriate acid for use in an analytical procedure without having to look up the acid dissociation constant every time.

Many factors affect acid strengths, including the identity and electronegativity of the atoms in the compound, the number of atoms in the molecule, the shape of the molecule, and the size of the atoms. In general, factors that contribute to stabilizing the products (H_3O^+ and A^-) relative to reactants (HA and H_2O) enhance acidity.

Two major factors help stabilize the conjugate base relative to the acid:

*   Delocalization of the negative charge on the conjugate base over a large area to reduce the charge density.

*   Stabilization of the negative charge by having it in close proximity to a strong positive charge (e.g. the nucleus of an electronegative atom).

In this Exploration, you will examine a series of acidic compounds, with the goal of answering the question *How is acid strength related to the structure and composition of the acid?*

How can you find out more?

The main group elements (B, C, N, P, S, As, F, Cl, Br, I) form a series of compounds called the **oxyacids** that result from the combination of the central atom with 2-4 oxygen atoms. The oxyacids have a range of acidities and provide an excellent way to examine the relationship between acid strength, molecular structure, and the stability of the conjugate base.

Developing Ideas

The structures and K_a values of some of the oxyacids are shown below. Use these structures as a starting point for doing the exercise.

Oxyacid	Structure	K_a
H_3BO_3	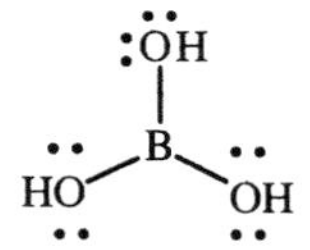	7.3×10^{-10}
H_2CO_3	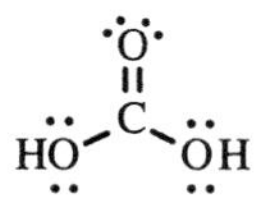	4.3×10^{-7}
HNO_3	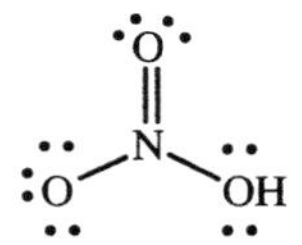	~10
HNO_2	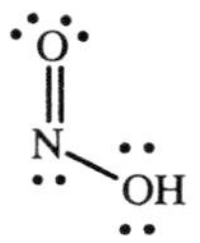	4.6×10^{-4}
H_2SO_4	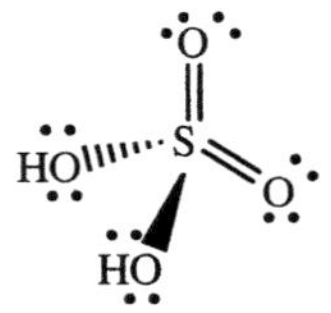	~1000
H_2SO_3	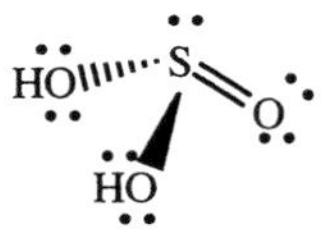	1.54×10^{-2}
H_3PO_4	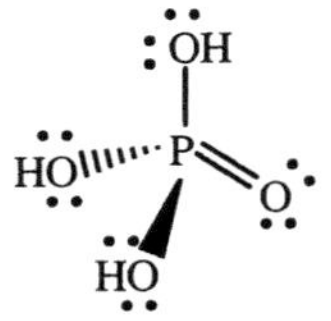	0.008
$HClO_4$	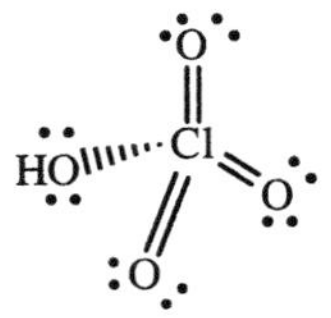	~10^7
HF		3.5×10^{-4}
HCl		~1000
HBr		~5000
HI		~10,000

1.   Compare the acidity of the oxyacids within the groups shown.

Group 1: H_3BO_3, H_2CO_3, HNO_3

Group 2: H_2SO_4, H_3PO_4, $HClO_4$

Pauling Electronegativities

B	2.0	S	2.5
C	2.5	P	2.1
N	3.0	Cl	3.0
F	4.0	Br	2.8

a.   Using the K_a values, rank the compounds from most acidic to least acidic.

b.   Examine the central atom in each of the oxyacids. Using the table of electronegativities given in the side-bar, rank the central atoms from *least electronegative* to *most electronegative*.

c.   How is K_a for these oxyacids affected by the electronegativity of the central atom?

d.   Generalize this result. How is the acid strength of the oxyacids affected by the electronegativity of the central atom?

e.   Explain why your generalization is true by commenting on the relative stabilities of (H_2O + acid) and (H_3O^+ + conjugate base).

2.   Compare the acidity of the following pairs of acids by looking up their K_a's.

H_2SO_4 and H_2SO_3

HNO_2 and HNO_3

a.   What generalization can you make about the relationship of the number of oxygen atoms in the oxyacid to acid strength?

b.   Draw all possible resonance structures for the conjugate base of each acid.

c.   Why does increasing the number of oxygen atoms stabilize the conjugate base relative to the acid?

3.   Compare the K_a values of HF, HCl, HBr, and HI. The sizes of the F^-, Cl^-, Br^-, and I^- ions are 1.17, 1.67, 1.82, and 2.06 Å, respectively.

a.   What generalization can you make about the relationship of the size of the counterion to acid strength? Explain why this is true.

b.   Which of these acids would you predict would be the strongest based on the electronegativity of the halogen atom? Explain.

Working with the Ideas

The following problems will help you understand the concepts in more depth.

4.   Which compound in each of the following pairs would you expect to be more acidic? Draw Lewis dot structures and accurate representations of the molecular shapes and use resonance structures to explain your reasoning.

a. H_2SO_3 or H_2SO_4

b. HSO_3F or H_2SO_4

c. H_3PO_4 or $H_2PO_4^-$

d. $HClO_4$ or HIO_4

e. CH_3COOH or CCl_3COOH

f.   HCN vs. HCl

g. H_2S vs. H_2O

Making the Link

Looking Back: What have you learned?

Chemical Principles

What chemistry experience have you gained?

As you worked through Session 7, you gained experience with some principles of chemistry that can help solve a wide range of real-world problems. Your experience with Session 7 should have developed your skills to do the following items.

Acids and Bases

You should be familiar with:

- The concept that acids (HA) react with water, dissociating to make H_3O^+ ions and A^- ions. The greater the extent of the reaction, the stronger the base (**Explorations 7A and 7B**).

- The concept that bases (B) react with water to make OH^- ions and BH^+ ions. The greater the extent of the reaction, the stronger the base (**Explorations 7A and 7B**).

- The definition of an acid as a proton donor and a base as a proton acceptor (**Exploration 7B**).

- The concept that water is both an acid and a base (**Exploration 7B**).

- The concept that the stronger the acid, the weaker the conjugate base (**Exploration 7B**).

- The definition of a conjugate acid as the weaker acid resulting from the reaction of a primary acid with water (**Exploration 7B**).

- The definition of a conjugate base as the weaker base resulting from the reaction of a primary base with water (**Exploration 7B**).

- The use of K_a to represent the equilibrium constant for the ionization of an acid and K_b to represent the equilibrium constant for the ionization of a base. An acid with a large K_a is strong and one with a small K_a is weak. Similarly, a base with a large K_b is strong and one with a small K_b is weak (**Exploration 7B**).

- The use of K_w to represent the equilibrium constant for the ionization of water (**Exploration 7B**).

Structure-Acidity Relationships

You should be able to:

- Determine whether a compound is acidic or basic based on its structure and composition (**Exploration 7A**).

- Discuss the factors that affect the acidity of the compound. In general, factors that contribute to the stability of the conjugate base favor the H_3O^+/A^- pair relative to the HA/H_2O pair. Delocalization of charge on the conjugate base stabilizes it relative to the acid. For oxyacids, the electronegativity of the central atom and the number of electronegative atoms bound to the central atom play a major role in increasing acidity. For the hydrohalic acids, HX, the size of the conjugate base anion, X^-, is the most important factor in determining acidity, overriding the importance of electronegativity (**Exploration 7F**).

Equilibrium Expressions Involving Acids and Bases

You should know:

- How to write equilibrium expressions for acid-base reactions (**Exploration 7B**). For the reaction

$$HA \;+\; H_2O \;\overset{K_a}{\rightleftharpoons}\; H_3O^+ \;+\; A^-$$

    acid   base   conjugate   conjugate
                                   acid   base

The equilibrium expression is written as:

$$K_a = \frac{\left[H_3O^+\right]\left[A^-\right]}{\left[HA\right]}$$

where the concentration of the water in the reaction is a constant and is therefore incorporated into the equilibrium constant.

- The value of the acid dissociation constant provides information on the extent of the reaction to form H_3O^+ and A^-. If K_a is greater than 1, the acid is classified as **strong** and the dissociation reaction proceeds to a large extent. If K_a is less than 1, the acid is not as extensively ionized and may be classified as a **weak** acid (**Explorations 7B** and **7D**).

The pH Scale

You should be familiar with:

- The two scales for reporting acid and base concentrations, mol/L and pH, where pH is -log[H_3O^+]. The log scale provides information on the order of magnitude of a concentration, so when concentration differences are large, the numbers stay manageable (**Exploration 7C**).
- The interrelationship of OH^- and H_3O^+ and how to use the relationship to find out the concentration of both species in solution (**Explorations 7B** and **7C**).

Acid-Base Equilibrium Calculations

You should know:

- How to predict equilibrium concentrations using the equilibrium expression, the acid or base ionization constant, and knowledge of the stoichiometry of the reaction (**Exploration 7D**).
- When you can make the assumption that the amount of a weak acid that is dissociated is small relative to the concentration (**Exploration 7D**). Typically, if the concentration is different from the K_a value by three orders of magnitude or more, the amount dissociated can be ignored. This simplifies the equilibrium expression to a system that can be solved without using the quadratic formula.

Thinking Skills

What general skills are you building for your resume?

You have also been developing some general problem-solving and scientific thinking skills that are valued by employers in a wide range of professions and in academia. Here is a list of the skills that you have been building for your resume.

Data Analysis Skills

You should be able to:

- Plot a data set and interpret the plot to reach a conclusion about a system (**Exploration 7E**).

Mathematical Skills

You should be able to:

- Use logarithms to interpret and explain data that differ by orders of magnitude (**Exploration 7C**).

Analogical and Deductive Reasoning Skills

You should be able to:

- Gather and organize data to look for trends and differences (**Explorations 7A and 7B**).
- Apply a set of principles learned in one setting to make predictions in a different setting (**Explorations 7A and 7B**).
- Use existing data to draw analogies to a new situation (**Explorations 7A, 7B, and 7E**).

Checking Your Progress

What progress have you made toward answering the Module Question?

Appendix 7A

Acid and Base Ionization Constants

Table 7-3   Acid Ionization Constants for Strong Acids, pK_a, at 25°C

Acid	Structure	Conjugate Base	pK_a
Hydrochloric acid, HCl			-3
Nitric acid, HNO_3			0
Perchloric acid, $HClO_4$			-7
Sulfuric acid, H_2SO_4			-3
Phosphoric acid, H_3PO_4 (intermediate acid)			2.1

Table 7-4   Acid Ionization Constants for Weak Acids, K_a, at 25°C

Acid	Structure	Acid Ionization Constant, K_a	pK_a
Acetic acid, CH_3COOH		1.8×10^{-5}	4.7
Bisulfate ion, HSO_4^-		1.3×10^{-2}	1.9
Carbonic acid, H_2CO_3		4.2×10^{-7}	6.4
Citric acid, $C_6H_8O_7$		7.4×10^{-4}	3.1
Formic acid, $HCOOH$		1.7×10^{-4}	3.77
Hydrocyanic acid, HCN	$H — C \equiv N:$	4.9×10^{-10}	9.3
Hydrofluoric acid, HF	$H — \overset{..}{\underset{..}{F}}:$	7.1×10^{-4}	3.2
Hypochlorous acid, $HOCl$		2.5×10^{-8}	7.6
Phenol, C_6H_5OH		1.3×10^{-10}	9.9
Water, H_2O		1.0×10^{-14}	14.0

Table 7-5   Protonation Constants for Weak Bases, K_b, at 25°C

Base	Structure	Base Protonation Constant, K_b	pK_b
Ammonia, NH_3		1.75×10^{-5}	4.76
Bicarbonate, HCO_3^-		2.0×10^{-8}	7.7
Carbonate, CO_3^{2-}		2.0×10^{-4}	3.70
Methyl amine, NH_2CH_3		4.59×10^{-4}	3.34
Pyridine, C_5H_5N		1.46×10^{-9}	8.84
Water, H_2O		1.0×10^{-14}	14.0

Session 8 Acids and Bases II

What is the role of acids and bases in water treatment?

Exploration 8A: The Storyline

Creating the Context

Why is this an important question?

Humans and animals can tolerate fairly large extremes in the pH of their drinking water. At one extreme, most soft drinks have a pH between 2 and 4! Raw (untreated) public water supplies typically have a pH between 4 and 9, with the majority having a pH between 5.5 and 8.6. After treatment, most public water supplies have a pH between 6.9 and 7.4. The acceptable pH range for drinking water as dictated by the Public Health Service Act is 6.5-8.5.

As you may have noticed in your laboratory work in Session 6, many of the substances used as water treatment reagents are acidic or basic. In fact, acids and bases play critical roles in many areas of water treatment chemistry. In this Session, we will examine how pH can alter the solubility of substances both beneÞcial and toxic. The goal of this Session is to answer the question, *What is the role of acids and bases in water treatment?*

Developing Ideas

Your instructor will demonstrate the effects of changing the pH of several different solutions. Observe the demonstrations and note any changes in the reaction mixture.

Acid added to:	Indication(s) of reaction?	Add KI^* Observation?
Pb shot		
Cu wire		
Brass nut, Sn, Pb, Cu		
Dolomite, $CaMg(CO_3)$		
Base added to:		
Solution from Pb shot plus acid		
Solution from Cu wire plus acid		
Solution from brass nut plus acid		
Hard water (high Mg^{2+}/Ca^{2+})		

*Potassium iodide, KI, serves as an indicator of the presence of metal ions, precipitating out colorful metal iodide salts, MI_2 in the reaction: M^{2+} (aq) + 2 I^- (aq) ----> MI_2 (s).

1. Water pipes are frequently made of copper, brass, and, in the past, lead. What problems might arise if a water supply is too acidic?

2. What problems might arise if a water supply is too basic?

3. Under which conditions would you expect to see a higher concentration of dissolved substances, high pH or low pH? Explain your reasoning.

4. Why do you think the water quality standards specify that the pH of a drinking water supply be between 6.5 and 8.5?

Looking Ahead

Several Exploration Questions can guide you to explore the role of acids and bases in water treatment in more depth. These will be used at the discretion of your instructor.

- **Exploration 8B:** How can we selectively remove ions from solution?
- **Exploration 8C:** How do you measure pH?
- **Exploration 8D:** How do you change the pH of a water supply?

Exploration 8B

How can we selectively remove ions from solution?

Creating the Context

Why is this an important question?

When treating raw water to remove dissolved minerals, water treatment engineers often raise the pH to precipitate out heavy metals and water hardness ions. These species form nearly insoluble complexes with a base like hydroxide, OH^-, and can be removed essentially quantitatively. In this process, you must add the correct amount of base. If you add too much, the pH of the water will be higher than needed to remove the contaminants, and more reagent will be used up than required. In addition, more acid will be required to bring the pH back within range of the water quality standards. If you add too little base, you may not successfully remove all of the contaminants.

The control of pH is also important in the analysis of water quality. For example, both the iron and water hardness analyses require that the pH be around 10. The hardness analysis provides a particularly good example of selective removal of ions from a solution. In this Exploration, we will examine the process by which ions can be selectively precipitated by controlling the pH of the solution. By the end of this Exploration, you should be able to answer the question *How can we selectively remove ions from solution?*

Developing Ideas

You wish to determine water hardness due only to calcium ions in solution using an EDTA titration (see Exploration 2C). Typical concentrations of the hardness ions Ca^{2+} and Mg^{2+} in natural waters range from 0.0005 to 0.005 M. One approach is to remove one of the ions by precipitation, then do the analysis and compare it to the analysis of the sample without precipitation. How can you determine the pH at which the analysis must be carried out to analyze for *only* the Ca^{2+}?

1. Begin by writing the solubility product expressions for $Ca(OH)_2$ and $Mg(OH)_2$, looking up the K_{sp} values in Appendix 4A.

2. Calculate the pH at which Mg^{2+} will begin to precipitate out of solution by determining the $[OH^-]$ at which $Q = K_{sp}$ for both ends of the range of expected Mg^{2+} concentration.

3. Calculate the pH at which Ca^{2+} will begin to precipitate out of solution by determining the $[OH^-]$ at which $Q = K_{sp}$ for both ends of the range of expected Ca^{2+} concentration.

4. Use your results to determine the optimum pH for the analysis.

Working with the Ideas

5. A water source near a mining area contains 100 mg/L of Ni^{2+} and 20 mg/L of Pb^{2+}. If the pH is raised to 8.5, will enough of the metals be removed so the concentration of both of these metals drops below the MCL? For MCLs, see the Environmental Protection Agency's web site at http://www.epa.gov/water-science/drinking/

6. Waste water from an electroplating facility contains Cr^{3+}, Ni^{2+}, and Cu^{2+}. The pH is raised to precipitate the metals.

   a. Which metal will precipitate out of solution at the lowest pH?

   b. Which metal will remain in solution the longest?

   c. How could you use pH adjustment to purify an electroplating solution so it contained only one metal ion?

Exploration 8C

How do you measure the pH of a water supply?

Creating the Context

Why is this an important question?

What background information is useful?

In Exploration 8A, you learned the importance of controlling pH to minimize the concentrations of toxic metal ions and water hardness ions in a water supply. If we wish to control pH, we need to be able to monitor pH changes in solution. In this Exploration, you will learn the theory and practice behind measuring pH in the laboratory. You have encountered a similar analytical method if you used a ßuoride-selective electrode (see page 26) to analyze for ßuoride.

Most laboratory pH measurements are carried out using a pH meter. The pH meter works by using an electrical circuit that contains a glass electrode, an external reference electrode, and a voltmeter that measures the electrical potential in the circuit. The glass electrode consists of a current-carrying wire that dips into a solution of known pH (see Figure 8-1). This solution is encapsulated in a thin glass membrane made of aluminosilicate glass (see Figure 8-2). The molecular structure of the glass is such that oxygen atoms comprise much of the surface of the glass, with the backbone of the molecule consisting of aluminum and silicon atoms. Because oxygen atoms are electronegative and have unshared pairs of electrons, they will have a partial negative charge and will thus be "sticky" to positively-charged species like protons. When the glass electrode is dipped into an acidic sample to measure the pH, the protons in the sample stick to the outside of the glass membrane. To maintain neutrality of the membrane, some of the protons that are stuck to the inside surface of the glass membrane desorb from the surface. The consequence is a change in the pH of the inner solution, which results in an electrical potential.

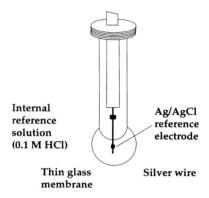

Figure 8-1: The glass electrode, used for measuring pH.

When the electrode is dipped into a basic sample, the protons adhering to the external surface of the glass electrode are removed by the hydroxide ions in solution. Again, to neutralize the charge on the membrane, protons from the inner solution adhere to the inner surface, thereby changing the electrical potential within the electrode. A reference electrode is coupled to the glass electrode to complete the circuit. The voltmeter measures the electrochemical potential of the cell, which is then translated into a pH reading by the meter.

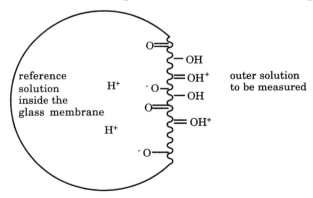

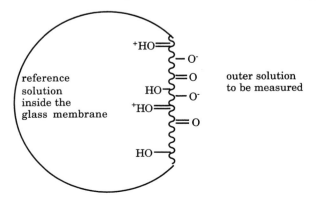

Figure 8-2: The surface of the glass electrode in acidic and basic solution.

Laboratory Procedures

Calibration of the pH Meter

To obtain accurate readings, the pH meter must Þrst be calibrated using solutions of known pH. Buffer solutions of known pH work well for this purpose. Calibration buffers are chosen with pH values that bracket the expected pH of the sample solution, usually pH 4.0 and 7.0 or 7.0 and 10.0. Although the exact instructions vary slightly by model, the general standardization procedure is outlined below.

---

note   All glassware used in this experiment should be very clean and well rinsed with distilled water. If the glassware will not come clean with normal procedures, use the procedures in Appendix 2A.

The standard procedure for transferring the electrode from one solution to another is as follows: Remove the electrode from the solution and rinse it *well* with deionized water, using your wash bottle. Shake the excess water off the electrode and place it in the new solution.

---

note   Do not touch the bottom of the electrode, as this will depolarize it and give inaccurate readings.

---

note   Do not let the electrode dry out. Be sure it is immersed in a solution except when transferring between beakers.

---

1. Turn the meter on. Rinse the electrode with de-ionized water, shake it dry, and place it in a beaker containing the Þrst buffer solution at pH 7.00. If you are using a magnetic stirrer, make sure a magnetic stir bar is in the beaker and adjust the stirrer so the stir bar spins slowly. Wait for the reading to stabilize, then adjust the reading to pH 7.00.

2.   Rinse the electrode with de-ionized water, shake it dry, and place it in a beaker of the second standard buffer solution (pH 4.00 or 10.01, depending on the expected pH of the samples). Allow the reading to stabilize and then adjust it to the appropriate pH for the buffer solution used.

3. Check the calibration by measuring the pH of the 7.00 standard. To do so, place the electrode in the solution, wait for a stable reading, and note the number. If the reading is more than 0.05 pH units different from 7.00, repeat the standardization procedure, steps 1 through 3).

4. At this point the meter is ready to use.

Measurement of Sample pH Using a pH Meter

---

note   All glassware used in this experiment should be very clean and well rinsed with distilled water. If the glassware will not come clean with normal procedures, use the procedures in Appendix 2A.

Take the pH of your sample by Þlling a scintillation vial with enough sample to immerse the (clean) electrode into the vial. Stir the solution, allow the reading to stabilize, and read the display to obtain the pH of your sample. Record this value in your notebook.

If you dispose of your sample at this point, you have not added any hazardous substances, so it may be disposed of down the drain.

disposal

Exploration 8D

How do you change the pH of a water supply?

<table>
<tr><td>

Creating the Context

Why is this an important question?

What do you already know?

Preparing for Inquiry

</td><td>

In Session 6, you spent several laboratory periods developing a method for removal of a contaminant from a water sample. Many of the reagents used to remove contaminants are acids or bases, and as such, will change the pH of the water sample. Because water supplies out of the range of pH 6.5-8.5 can be hazardous to health and plumbing Þxtures, the pH must be adjusted back to this range before distribution in the system. **Neutralization** is the process by which the pH is brought back to neutral and is central to all water treatment plants.

You should already know which reagents have the potential to make the sample acidic or basic, by remembering the characteristics of acidic and basic compounds. In this Exploration, you will design a plan for bringing the pH into the 6.5-8.5 range for your treated water sample. After this experiment, you should be able to answer the question *How do you change the pH of a water supply?*

</td></tr>
</table>

BACKGROUND READING

Adjusting pH During Water Treatment

The water treatment processes involving precipitation of hydroxide or carbonate salts are pH-dependent because the hydroxide and carbonate anions react with acid to form new compounds and remove the precipitating anion. Adjustment of pH during water treatment is thus critical to many processes that remove contaminants. For example, consider the dissolution of magnesium hydroxide, $Mg(OH)_2$ in pure water versus in water at pH 11.

$$Mg(OH)_2 \text{ (s)} \rightleftharpoons Mg^{2+}(aq) + 2\,OH^-\,(aq)$$

$$K_{sp} = \left[Mg^{2+}\right]\left[OH^-\right]^2 = 1.2 \times 10^{<11}$$

In pure water at 25°C, the concentration of magnesium can be determined using the solubility product expression:

$$s = \text{molar solubility}$$

$$1.2\times10^{-11} = (s)(2s)^2 = 4s^3$$

$$s = \sqrt{3.0\times10^{-12}} = 1.4\times10^{-4}\,\text{M}$$

$$1.4\times10^{-4}\,\text{M} \times \frac{24.3\ g}{\text{mol}} \times \frac{1{,}000\ mg}{g} = 3.4\ \text{mg/L (ppm)}$$

At pH 11.0 and 25°C, the [OH$^-$] = 2.0 x 10^{-3} M, and the calculation gives a very different result:

$$1.2\times10^{-11} = (s)(2.0\times10^{-3})^2$$

$$s = \frac{1.2\times10^{-11}}{4.0\times10^{-6}} = 1.2\times10^{-5}\ \text{M}$$

$$1.2\times10^{-5}\ \text{M} \times \frac{24.3\ g}{\text{mol}} \times \frac{1{,}000\ mg}{g} = 0.29\ \text{mg/L (ppm)}$$

Note how much *less* magnesium remains in solution at pH 11.0! The clever person who Þgured out that unwanted ions could be precipitated out by raising the pH must have been pretty proud of that result!

In many water treatment processes, the pH of the water supply is raised, usually to between 10 and 11, to cause unwanted ionic species to precipitate out of solution. After the precipitation reaction is complete, the excess base must be neutralized to bring the pH back into the acceptable range of 6.5-8.5.

Many chemicals can adjust the pH of water to a desired value. Table 8-1 lists the chemicals available in the laboratory for adjusting the pH of your water sample. Part of your job will be to determine which is the most convenient and cost-effective but adds the least amount of additional ionic substances to the water supply.

Table 8-1   Chemicals Used for Adjustment of pH in Water Treatment

Adjusts pH Up	Equilibrium Constant	Adjusts pH Down	Equilibrium Constant
0.005 M NaOH	K_b = very large	0.005 M HCl	K_a = very large
0.005 M KOH	K_b = very large	0.005 M H_2SO_4	K_{a1} = very large
			K_{a2} = 1.3x10^{-2}
$Mg(OH)_2$ (s)	K_{sp} = 1.2x10^{-11}	0.005 M NaHSO$_4$	K_a(HSO_4^-) = 1.3x10^{-2}
$Ca(OH)_2$ (s)	K_{sp} = 8.0x10^{-6}	CO_2 (g)	K_a (H_2CO_3) = 5.0x10^{-7}
			K_a (HCO_3^-) = 5.0x10^{-11}

Compounds Used to Raise pH

Most of the Group I and II hydroxides and carbonates are effective bases for water treatment processes. The Group I (Na, K) hydroxides and carbonates are extremely soluble in water. The Group II (Ca, Mg) hydroxides and carbonates are only partially soluble, an attribute that can be both desirable (if you do not want to add too much of the counterion to the water supply) and undesirable, because it might not be possible to dissolve enough of the compound to raise the pH to the desired level.

In this experiment, you will calculate the amount of solid or concentrated solution of these compounds required to adjust the pH of your water sample to the desired value.

Calculations for the hydroxides involve only one equilibrium, that of dissolution of the hydroxide salt. Some example calculations follow.

Example 1

How many grams of solid NaOH would be required to raise the pH of 1000 liters of pure water at 25°C in a reactor from neutral to pH 11.0? What volume of 6 M NaOH would be required?

Answer 1

Begin by finding [OH⁻] of the final solution in moles per liter. We will assume $T = 25°C$, where $K_w = 1 \times 10^{-14}$. At this temperature

$$pH + pOH = 14, \text{ so pOH} = 3.0 \text{ and } [OH^-] = 10^{-3.0} = 1 \times 10^{-3} \text{ M}$$

Now use the molecular weight of NaOH and the volume of water to be treated to find the amount of solid NaOH required.

$$\frac{1 \times 10^{-3} \text{ moles}}{L} \times \frac{40 \text{ g}}{\text{mole}} \times 1{,}000 \text{ L} = 40 \text{ g of NaOH}$$

If the NaOH is available as a concentrated (6 M) solution, the calculation is slightly different. First, find the number of moles of NaOH, then calculate the volume of the 6 M NaOH solution required to reach the desired pH.

$$\frac{1 \times 10^{-3} \text{ moles}}{L} \times 1{,}000 \text{ L} = 1 \text{ mol}$$

$$1 \text{ mol} \times \frac{1 \text{ L}}{6 \text{ mol}} = 0.17 \text{ L}$$

Example 2

What would be the resulting concentration of sodium ions in mg/L (ppm) for the solution above?

Answer 2

$$\frac{1 \times 10^{-3} \text{ moles}}{L} \times \frac{23 \text{ g}}{\text{mole}} = \frac{0.023 \text{ g}}{L} = \frac{23 \text{ mg}}{L}$$

Example 3

What is the highest pH that can be achieved using solid $Mg(OH)_2$, $K_{sp} = 1.2 \times 10^{-11}$, as the base at 25°C?

Answer 3

The concentration of hydroxide ion that can be achieved using $Mg(OH)_2$ as a base is limited by the amount of solid that will dissolve, according to the K_{sp}.

$$s = \text{molar solubility}$$

$$1.2 \times 10^{-11} = \left[Mg^{2+} \right]\left[OH^- \right]^2 = (s)(2s)^2 = 4s^3$$

$$s = 1.4 \times 10^{-4} \text{ M}; \left[OH^- \right] = 2s = 2.8 \times 10^{-4} \text{ M}$$

$$pOH = -\log\left(2.8 \times 10^{-4} \text{ M}\right) = 3.6; \ pH = 14 < 3.6 = 10.4$$

Lowering pH with Strong Acids

A variety of inexpensive strong acids are commonly used in water treatment plants, including hydrochloric, nitric, and sulfuric acids. These chemicals are usually called into action to neutralize the excess base added to remove contaminants by precipitation at high pH.

Example

A water treatment engineer has a 1000 L batch reactor of water that is at pH 10.4. What volume of 12 M HCl must be added to the water supply to bring the pH down to 8.5, the highest allowable pH?

Answer

To bring the pH in range of the water quality standards, enough acid must be added to neutralize the base present. Begin the problem by determining the initial $[HO^-]$ and the desired final $[HO^-]$.

Initial $[HO^-]$: At pH 10.4, pOH = 3.6 and $[HO^-] = 2.51 \times 10^{-4}$ M

Final $[HO^-]$: At pH 8.5, pOH = 5.5 and $[HO^-] = 3.16 \times 10^{-6}$ M

The difference between these two, adjusted for the volume of water, gives the amount of base that must be neutralized:

$(2.51 \times 10^{-4}$ M$) - (3.16 \times 10^{-6}$ M$) = 2.48 \times 10^{-4}$ M

$(2.48 \times 10^{-4}$ M$) \times 1000$ L $= 0.248$ moles of OH$^-$ to be neutralized

Since HCl has one acidic proton, the moles of acid required will be the same as the number of moles of base to be neutralized. Using the concentration of the HCl as 12 M, the volume of HCl required to bring the pH to 8.5 can be calculated:

$$0.248 \text{ mol} \times \frac{1 \text{ L}}{12 \text{ mol}} = 0.0207 \text{ L of HCl}$$

Lowering pH with Carbon Dioxide

A cheap way of lowering the pH of water without adding additional ions is to bubble CO_2 through the water. Carbon dioxide reacts with water to form carbonic acid, which neutralizes any base present, leaving the solution with a concentration of HCO_3^- equal to the concentration of base initially present.

$$CO_2 \text{ (g)} + 2 H_2O \text{ (l)} \longrightarrow H_2CO_3 \text{ (aq)}$$

$$H_2CO_3 \text{ (aq)} + HO^- \text{ (aq)} \longrightarrow HCO_3^- \text{ (aq)} + H_2O \text{ (l)}$$

---

note

A Caveat: In your laboratory work, you may find that the measured pH of a solution is not what you calculate it should be. The reasons might be one of the following:

The solution may not have reached equilibrium at the time of the measurement. Some of these reactions may take hours to reach

If there are high concentrations of other ions in the solution, the pH meter may give erroneous readings.

You might not have measured the proper amount of reagent.

Laboratory Procedures

The reagents available in the laboratory are listed in Table 8-1 on page 170. Using the general laboratory procedures shown below, experiment with these reagents to balance the pH of your water sample. Consider these factors when deciding which is the optimum method:

- Cost of the overall system
- Efﬁciency of the process
- Rate at which the water can be treated
- Simplicity of the overall system

Your goal is to optimize the process for bringing the pH into an acceptable range. The general approach requires you to:

- Calculate the expected amount of reagent required to adjust the pH of the treated sample.
- Add the appropriate reagent and measure the pH of the sample.
- Adjust the variable you are experimenting with and repeat the experiment until your pH is in the correct range.

> **note** Because you will be designing your own procedures, it is **essential** that you keep a *detailed, accurate* account of **everything** you do experimentally. If in doubt, write it down! Your ﬁnal report depends critically on your notes.

Adjusting the pH Upward

Calculate the amount of reagent required to bring the pH to the desired value for the volume of water being treated, then experiment to ﬁnd which one gives the best results for your water treatment plan. Calculate the cost of the pH adjustment process using the data in Appendix 6B on page 139. There are ~20 drops in a milliliter.

Adjusting the pH Downward

In addition to the solutions and solids in Table 8-1, CO_2 gas (dry ice) is also available. Calculate the amount of reagent required to bring the pH to the desired value for the volume of water being treated, then experiment to ﬁnd which one gives the best results for your water treatment plan. Calculate the cost of the pH adjustment process using the data in Appendix 6B on page 139. There are ~20 drops in a milliliter.

Using CO_2 to bring the pH down requires a special apparatus that uses dry ice to generate CO_2 gas. Figure 8-3 diagrams this apparatus, easily constructed from your laboratory glassware.

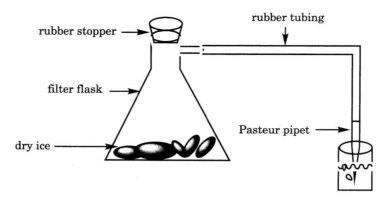

Figure 8-3:   Apparatus used to bubble carbon dioxide through a water sample.

1.   Attach one end of a piece of rubber tubing to a clean Pasteur pipet and the other end to a Þlter ßask that is clamped for stability.

2.   Place a few pieces of dry ice in the Þlter ßask and put the rubber stopper on the ßask to force the gas out the pipet tip.

3.   Place the pipet tip in the sample and bubble CO_2 through the solution until the desired pH is reached. Use pH paper as a quick check of the progress of the reaction.

Making the Link

Looking Back: What have you learned?

Chemical Principles

What chemistry experience have you gained?

In the process of working through Session 8, you have gained experience in working with some important principles and techniques of chemistry. An understanding of these principles and techniques is valuable in solving a wide range of real-world problems. Your experience with Session 8 should have developed your skills to do the items on the following lists.

Selective Precipitation

You should be able to:

•   Use K_{sp} to selectively remove one ion in the presence of others by adjusting the pH (**Exploration 8B**).

Neutralization

You should know

•   The concept of **neutralization**, where acid or base is added to a solution to bring the pH back to neutral (**Exploration 8D**).

- How to calculate the amount of reagent (a solid or a solution) required to neutralize a sample (**Exploration 8D**).
- How to recognize the general types of reagents used for neutralization processes (**Exploration 8D**).

Laboratory Measurements

You should be able to:

- Measure the pH of a solution (**Exploration 8C**).
- Understand how a pH meter works (**Exploration 8C**).

Thinking Skills

What general skills are you building for your resume?

In the process of working through Session 8, you have also been developing some general problem-solving and scientific thinking skills that are valued by employers in a wide range of professions and in academia.

Data Analysis Skills

You are able to:

- Use standards to ensure the accuracy of an analytical procedure (**Exploration 8C**).

Checking Your Progress

What progress have you made toward answering the Module Question?

Session 9 Projects

What are your results?

Now that we've reached the end of this module, you will present your work in a culminating project that will build on all the work you have done so far. Your instructor may assign one of the following projects or may ask you to choose among them.

Project 1: Poster Presentation

Project 2: Scientific Report

Project 3: Community Education

You already have most of the information you will need in your laboratory notebooks. Good luck!

Project 1: Poster Presentation

Scientists often present their data in poster format at professional meetings. Preparing a poster of your work in this module will introduce you to this way of sharing results with colleagues. Each team of students will produce one poster. Please use at most two pieces of poster board, available from local art stores.

A good poster makes it easy for people to understand what you did, why you did it, and what the results signify. Type all text, and include the following:

- Names of all students on the team and a descriptive title of your work.
- A brief introduction with background on your contaminant.
- A flow chart showing the sequence of experiments you used to optimize your treatment method.
- Data, in the form of tables, graphs, etc., showing the results of your experiments. Be sure to label all axes and use units.
- A summary of the chemical reaction(s) used in your treatment process.
- A brief description of the analytical methods you used to measure the contaminant concentration in your sample.
- Any other illustrative material to support your points (pictures, drawings, etc.)
- Conclusions: overall, what did you determine to be the optimum treatment process to remove your contaminant? Did you succeed in purifying your water? Do other problematic issues remain?

At the poster session:

Be ready to answer questions about the project, the techniques used, the chemical processes, and the calculations you did. The poster session will begin by having half of the teams stand with their posters while the other teams walk around to look at them. After about 30 minutes, the teams will switch.

Grading Scheme for Poster Session

Notebook pages: You will need to have your laboratory notebook pages available to be evaluated. They should be a complete record of your work.

Presentation and effort: Is the poster well organized? Does it look nice? Does it have all of the necessary components? Is there evidence that the team put some effort into solving the problem?

Ability to deal with questions: Can you answer questions about the following topics?

- How do the analytical methods you used in the experiment work? (Includes your contaminant, as well as TDS, pH, and alkalinity.)
- What are the chemical reactions responsible for removing your contaminant? Can you write the equations for those reactions and explain what is happening?
- Can you set up an equilibrium problem to solve? (Includes solubility equilibria and acid-base equilibria.)
- What experiments would you like to try to optimize your treatment plan further?
- What are the advantages and disadvantages of your treatment method?

Project 2: Scientific Report

A scientific report presents the results of a scientific study, describes the procedures used, and discusses the importance of the work in the context of solving a problem, or in the context of work that has been done before. A particular style is associated with scientific reports. The scientific style of writing attempts to convey what was done using the least number of words with the greatest clarity. While it may not be as exciting as the style you might use for an English paper, it focuses on the study, not the researchers, and contributes to whether the reader will take you seriously.

The report should include the following parts and be about 5-10 pages long, typed, double spaced, 1-inch margins, 12-point font.

I.   Introduction
II. Experimental Procedures and Results
III. Calculations
IV. Discussion and Conclusions
V.   Appendices
VI. References

STYLE: TENSE, VOICE, AND PERSON

In general, use past tense, passive voice, third person to describe what was done. Remember, you are describing *what you did*, not giving instructions to someone else. For example:

Correct: "The fluoride electrode was calibrated using a set of four standards."

Incorrect: "Calibrate the fluoride electrode by using a set of four standards."

Correct: "The water samples were collected at Lake Temescal in Oakland."
Incorrect: "While out cruising, we collected the water samples at Lake Temescal."

Correct: "The sample was dropped and, as a result, no data were obtained."
Incorrect: "I dropped the sample all over the floor and couldn't do any of the tests."

The incidence of I's, we's, you's, our's, your's, and my's should be very limited in a lab report. Instead, use passive voice, for example.

Correct: "The best procedure for removing excess fluoride from water was found to be adsorption onto activated charcoal."

Incorrect: "I found that activated charcoal was the best method for removing excess fluoride from water."

Avoid sensationalism. Back up any judgments with data and scientific facts and assess carefully whether you have enough data to come to a conclusion. The conclusion "Further study is needed" is also a valid one.

CONTENT

I. Introduction

The introduction should provide some background information on the ionic constituents of natural waters, the Safe Drinking Water Act, and a discussion of the problems encountered when a water supply exceeds the MCLs for pH, TDS, fluoride, water hardness, and iron. Set the stage for the study you carried out by letting the reader know the goals of this experiment. Include a brief description of the theory behind the methods used to remove the contaminant ion with which you worked.

II. Experimental Procedures and Results

The experimental section of your report should contain two subsections:
Section A: provides information about the unknown sample you started with.
Section B: details the experiments you carried out in an attempt to remove the contaminant.

Section A should contain a summary of the analyses of the initial unknown sample, referencing the procedures in the laboratory manual and summarizing the results of the initial analyses (pH, alkalinity, conductivity, and the concentration of your particular contaminant ion) in a table. *Don't forget to include your sample number.* You do not need to spell out the analytical procedures in detail, UNLESS you did something significantly different from the lab manual instructions.

Section B will describe the experiments you carried out to determine the best method to remove the contaminant ion. Since you did not have exact procedures in the laboratory manual for the water treatment experiments, you will need to rely heavily on the notes you took in your laboratory notebook. Use tables where possible to summarize results. A sample paragraph from an experimental procedure is shown below:

"Removal of excess fluoride was attempted as follows:
Experiment #1: Using a batch reaction scheme, 10 mL of sample F-2 was treated with 0.15 g of powdered charcoal. The mixture was stirred for 2.0 min-

utes and filtered through a syringe filter packed with glass wool and sand. The sample was then analyzed for fluoride content with a fluoride selective electrode and found to contain 4.5 mg/L of fluoride. The contact time was varied, keeping the amount of sample and charcoal constant. The filtration and analysis were repeated, with results shown in the table below:

Trial	Contact time (minutes)	Final F⁻ conc. (mg/L)
------	0 (original sample)	8.2
#1	2.0	4.5
#2	5.0	3.2
#3	10.0	1.8

The final pH of the Trial #3 sample was 5.2. Addition of 2 drops of 0.5 M NaOH to 10 mL of sample brought the sample into the 6.5-8.5 pH range required by the Drinking Water Standards."

III. Calculations

Show your work for any calculations. Include any graphs you created.

IV. Discussion and Conclusions

In this section, the results from the entire group should be presented and the pros and cons of the various remediation methods the group attempted should be discussed. The discussion should focus on the following:

- An overview of the experiments attempted by the group. Be sure to include the chemical equations of the reactions used to remove contaminants from the water sample.
- The results obtained from the experiments carried out.
- The preferred method for remediation found by the group.
- Other experiments you would like to do if there were more time.

Finally, answer the **Working with the Ideas** questions found at the end of either 6B (fluoride), 6C (hardness), or 6D (iron) in the lab manual. Include an estimate of the cost of your method for remediating 1 million gallons of water.

V. Appendices

Include one of the appendices below, according to the contaminant ion you worked on:

a. A discussion in your own words of the analytical method used to determine water hardness. Be sure to include the chemical equation that describes the reaction.

b. A discussion in your own words of the analytical method used to determine iron in water. Be sure to include the chemical equation that describes the reaction.

c. A discussion in your own words of the analytical method used to determine fluoride in water.

VI. References

It is very important to acknowledge all sources from which information was obtained in writing any scientific paper. The module can serve as your primary source material, but you might also wish to seek out and use other books or texts. A proper reference for a book includes the name of the author(s), the title, the publisher, the year, and the place of publication. For a journal article, the reference should include the name of the author(s), the journal title (in italics), the volume number, and the number of the page on which the article begins.

Project 3: Community Education

If your lab work involved studying a local water supply, your results may be of more than passing interest to those who live in your community. Your instructor may ask your class to present the results to local citizens groups, school children, or environmental groups.

Preparing an oral presentation requires organizing the information you wish to talk about and preparing visual aids, such as transparencies, to use during your presentation. Transparencies work better than notes because they give the audience something to look at and you don't have to look down to get your cues. Remember to think about the level of preparation of your audience and be prepared to spend a little more time explaining the scientific aspects of your project to a non-scientific audience.

You should include the following information in your transparencies:

- A title transparency with the names of all students on the team and a descriptive title of your work.

- A brief introduction to the location of the water supply and potential problems that might be encountered with water purity at that site.

- Background information on your contaminant.

- Data, in the form of tables, graphs, etc., showing the results of your experiments. Be sure to label all axes and use units.

- A summary of the chemical reaction(s) used in your treatment process. You may wish to omit this level of detail if you are speaking to a non-scientific audience.

- A brief description of the analytical methods you used to measure the contaminant concentration in your sample.

- Any other illustrative material to support your points (pictures, drawings, etc.).

- Conclusions: Overall, what did you determine to be the optimum treatment process to remove your contaminant? Did you succeed in purifying your water? Do other problematic issues remain?

At the presentation:

Be ready to answer questions about the project, the techniques used, the chemical processes, and the site. You may also wish to bring a demonstration to show the audience how your water purification technique works.

Exploration 3D

Why are some minerals more water soluble than others?

In Exploration 3C, you classified a set of organic compounds into polar and non-polar categories and correlated that classification to their water solubility. The correlation allowed you to predict that polar compounds will dissolve in polar solvents like water and non-polar compounds will dissolve in non-polar solvents. However, several factors seemed to affect the solubility of the *ionic* compounds in Sets 1 and 3. Because naturally-occurring substances that leach into water supplies are often ionic, we need to find out more about the factors that control their water solubility. We can then begin to predict what constituents to expect in a water supply from a given geographical area. As you will see in Session 5, this knowledge can also be used to remove contaminants from a water supply.

You learned in Exploration 3C that dissolution is favored when ΔH_{solv} is negative. In this situation, the energy released from making new bonds (between solvent molecules and the solute) exceeds the energy required to break bonds (the hydrogen bonds between the water molecules) plus the intermolecular forces of the solute; see Figure 3-4). Enthalpy changes are only part of the story, however, and the extent of dissolution of a compound depends on the magnitude and sign of the free energy change, ΔG_{solv}, for the reaction. This requires that we also take a closer look at the effects of *entropy* changes, ΔS_{solv}, for the dissolution of an ionic solid.

Recall that ΔS provides information about the change in the disorder of the system as the reaction proceeds from reactants to products. The combination of enthalpy and entropy effects explains why some ionic compounds do not dissolve readily in water, even though they are extremely polar. In this Exploration, we will examine the factors that determine the strength of the bonds between ions in a crystal lattice. In the process, we will find the answer to the question *Why are some minerals more water soluble than others?*

Preparing for Inquiry

BACKGROUND READING

The Energetics of the Crystal Lattice

To make sense of trends in the solubilities of a variety of substances, we need to know more about the forces that keep ionic compounds together. A simple model is to view the positive and negative ions contained in the crystal lattice as charged spheres with a definable size (given as the radius of the ion). These oppositely-charged spheres are held together by strong electrostatic forces arising from attraction of unlike charges for each other.

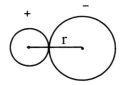

For two isolated gaseous ions, the energy of attraction is defined by Coulomb's Law,

$$E_{attraction} = \frac{kQ_1Q_2}{r}$$

where k is a constant, Q_1 and Q_2 are the charges on the ions and r is the internuclear distance between the ions. While this formula is helpful, it doesn't tell the entire story. In fact, an ionic solid is *more than* two isolated gaseous ions; it is an extensive lattice of positive and negative ions, and we must take all the attractions and repulsions among these ions into account in our bonding model (Figure 3D-1).

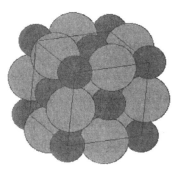

Figure 3D-1: Sodium chloride is an extensive repeating lattice of positive and negative ions. Each sodium ion is surrounded by six chloride ions (the larger, lighter-colored spheres) and each chloride ion is surrounded by six sodium ions (the smaller, dark spheres).

Enthalpy Considerations

The energy of attraction of a mole of ions in a crystal lattice is called the **lattice energy** or **lattice enthalpy** of the compound, ΔH_L, and is defined as the amount of energy required to transform one mole of an ionic solid into isolated gaseous ions (Note: Another symbol commonly used for lattice energy is U_0. The two are not precisely the same but are often used interchangeably). For example, the lattice energy of sodium chloride is defined as the energy required to break up the sodium chloride crystal lattice to form gaseous sodium and chloride ions. This is an extremely endo-

$$NaCl\ (s) \longrightarrow Na^+\ (g)\ +\ Cl^-\ (g) \qquad \Delta H_L = 786\ kJ/mol$$

thermic reaction, as you might expect for the highly unfavorable process of breaking apart positive and negative ions.

While this particular reaction rarely occurs under room temperature conditions, it is a convenient way to think of the process from a thermodynamic point of view, since the energies required to make positive and negative gaseous ions are relatively easy to measure, and tables of thermodynamic data can be consulted for these numbers. Table 3D-1 gives the lattice enthalpies for a variety of different compounds. The larger the lattice enthalpy, the stronger the attractive forces between ions in the crystal lattice.

Table 3D-1   Lattice Enthalpies of Representative Ionic Solids for the Reaction: SOLID---> IONS (g)

Compound	Lattice Enthalpy (kJ/mol)
Sodium fluoride, NaF	923
Sodium chloride, NaCl	786
Sodium bromide, NaBr	747
Sodium carbonate, Na_2CO_3	2,030
Sodium hydroxide, NaOH	887
Calcite, $CaCO_3$	2,810
Calcium chloride, $CaCl_2$	2,258
Calcium sulfate, $CaSO_4$	2,480
Calcium oxide, CaO	3,401
Calcium hydroxide, $Ca(OH)_2$	2,506
Magnesium oxide, MgO	3,791
Magnesium hydroxide, $Mg(OH)_2$	2,870
Iron (II) chloride, $FeCl_2$	2,631
Iron (III) chloride, $FeCl_3$	5,359
Iron (II) oxide, FeO	2,865
Hematite, Fe_2O_3	14,774
Quartz, SiO_2	13,125*
Zinc oxide, ZnO	4,142

*Calculated lattice enthalpy. All other numbers obtained from thermochemical data.
Source: *CRC Handbook of Chemistry and Physics,* 70th edition (CRC Press, Boca Raton, FL, 1989), pages D-104-D-110.

The entropy change for the disruption of a crystal lattice to form gaseous ions, ΔS_L, is positive, as you would expect for the transformation of one mole of a solid into two moles of gaseous ions. For salts composed of singly-charged ions (e.g., NaCl), the entropy change is in the range 200 to 300 J/mol·K. At 298K, $T\Delta S_L$ is on the order of 60 to 90 kJ/mol. The enthalpy term dominates the lattice energy reaction which results in the net free energy change for the disruption of a crystal lattice to form gaseous ions (ΔG_L) being positive, i.e. the reaction is strongly disfavored and non-spontaneous.

The Energetics of Dissolution

The ionic bonds of the crystal lattice are quite strong; however, for many ionic solids, the bonds between water molecules and dissolved ions are even stronger. In addition, the entropy change for the dissolution of many (but not all) ionic solids is positive, as the ions are released from the ordered crystal lattice to move more randomly in solution.[1] To understand the thermodynamics of the dissolution process, we must consider

---

1. When very small ions such as Li^+ and F^- dissolve, the entropy of the system actually *decreases* because these ions are very effective at ordering the water molecules in solution.

the energetics of the dissolution reaction, as indicated by the **free energy of solvation**, ΔG_{solv}. Remember that reaction spontaneity is governed by the magnitude and sign of the free energy of a reaction, ΔG, with a spontaneous reaction having a negative value of ΔG. With this in mind, we can say that in order for the dissolution of an ionic solid like NaCl, the free energy change for the dissolution reaction (ΔG_{solv}) must be negative.

The solvation or dissolution of an ionic solid can be viewed as the sum of two reactions (see Figure 3D-2).

1. The disruption of the crystal lattice to form gaseous ions, equivalent to the lattice energy reaction. The free energy associated with this process is ΔG_L.

$$NaCl\ (s) \longrightarrow Na^+\ (g) + Cl^-\ (g) \qquad \Delta G_L$$

2. The **hydration** of the gaseous ions to form aqueous ions. The free energy associated with this process is ΔG_{hyd}.

$$Na^+\ (g) + Cl^-\ (g) + x\ H_2O\ (l) \longrightarrow Na^+\ (aq) + Cl^-\ (aq) \qquad \Delta G_{hyd}$$

Important Reactions

Lattice energy: ΔG_L
solid >ions (g)

Hydration energy: ΔG_{hyd}
ions (g) >ions (aq)

Solvation energy: ΔG_{solv}
solid >ions (aq)

Both the hydration reaction and the lattice energy reaction rarely occur in the laboratory; however, breaking a reaction into elementary steps, even unlikely ones, helps to clarify the reactions for which we can calculate the energies.

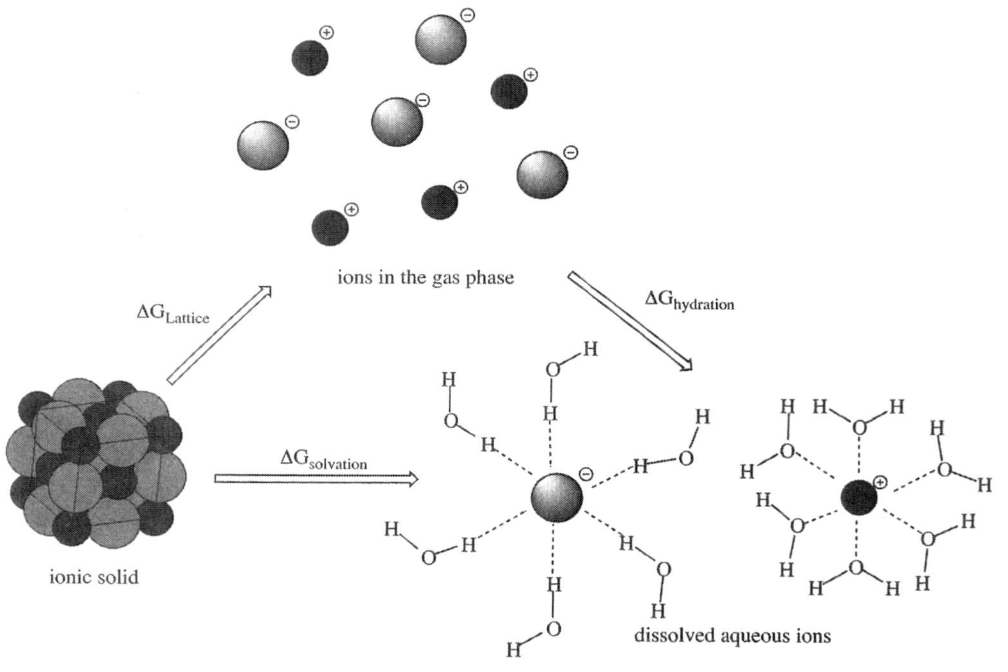

Figure 3D-2: Dissolution of an ionic solid can be viewed as the sum of two reactions: 1) the disruption of the crystal lattice to form gaseous ions, and 2) hydration of the gaseous ions to form aqueous ions.

How can we combine these two reactions and their associated free energies to chart the thermodynamics of the dissolution process? We must begin by expending energy to disrupt the crystal lattice and form gaseous ions, ΔG_L. These gaseous ions are then hydrated to form aqueous ions, a process that *releases* energy, ΔG_{hyd}. The free energy associated with the dissolution reaction, ΔG_{solv}, is thus the sum of the lattice energy and the free energy of hydration:

$$\Delta G_{solv} = \Delta G_L + \Delta G_{hyd}$$

For dissolution to occur, the free energy released on hydration of the gaseous ions must be greater than the free energy required to disrupt the crystal lattice and form gaseous ions, or more succinctly, ΔG_{solv} must be negative.

$$NaCl\ (s) \longrightarrow Na^+\ (g)\ +\ Cl^-\ (g) \qquad\qquad \Delta G_L$$

$$Na^+\ (g)\ +\ Cl^-\ (g)\ +\ x\ H_2O\ (l) \longrightarrow Na^+\ (aq)\ +\ Cl^-\ (aq) \qquad \Delta G_{hyd}$$

$$\overline{NaCl\ (s)\ +\ x\ H_2O\ (l) \longrightarrow Na^+\ (aq)\ +\ Cl^-\ (aq) \qquad\qquad \Delta G_{solv}}$$

Because $\Delta G = \Delta H - T\Delta S$, we can consider the relative contributions of enthalpy and entropy for the three reactions by plotting ΔH and $-T\Delta S$ for the three reactions on the same graph, as in Figure 3D-3, which summarizes the thermodynamic parameters for NaCl. (We obtained these values from Table 3D-2 and Table 3D-3.) On this plot, any energy that is negative contributes to reaction spontaneity.

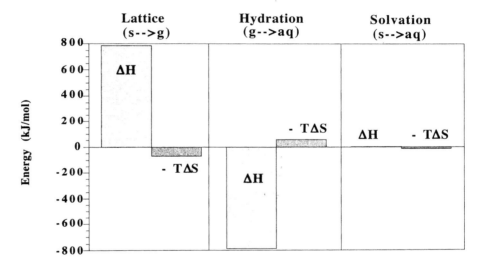

Figure 3D-3: The relative magnitudes of ΔH and $-T\Delta S$ terms for the lattice, hydration, and solvation reactions of NaCl are depicted on this plot. Negative bars contribute to the spontaneity of the dissolution reaction.

Let's examine the three parts of this graph. For the lattice reaction of NaCl, the free energy for the reaction to form gaseous ions from a solid, ΔG_L, is large and positive. While the entropy term ($-T\Delta S_L$) favors the reaction, its contribution is small and the dominant factor is the enthalpy contribution, ΔH_L.

For the hydration reaction of NaCl, the free energy is large and negative. The entropy term for the hydration reaction is small relative to the enthalpy term and does not favor the reaction since more order is introduced as gaseous ions are solvated by water molecules (see Figure 3D-2). Hydration enthalpies, ΔH_{hyd}, are *exothermic* because bonds are being formed between bare gaseous ions and water molecules. Because water molecules are polar, the process of hydration is much like the combination of positive and negative ions to form a lattice. As the oppositely-charged particles interact with each other, heat is released as bonds are formed. Note in Figure 3D-3 that the magnitude of the hydration enthalpy is very similar, but opposite in sign, to that of the lattice enthalpy, which tells us that the bonding interactions of an ion in a crystal lattice are very similar to those for a solvated ion in aqueous solution.

Table 3D-2 Hydration Entropies (ΔS_{hyd}) of Common Ions for the Reaction: ION (g) ----> ION (aq)

Ion	ΔS_{hyd} (J/mol·K)	TΔS (kJ/mol) at 25°C
Cations, M$^+$		
H$^+$	-109	-32
Li$^+$	-119	-35
Na$^+$	-89	-27
K$^+$	-51	-15
Rb$^+$	-40	-12
Cs$^+$	-37	-11
Ag$^+$	-94	-28
Anions, X$^-$		
F$^-$	-160	-48
Cl$^-$	-96	-29
Br$^-$	-80	-24
I$^-$	-62	-18

Source: W. E. Dasent, *Inorganic Energetics*, 2nd edition (Cambridge University Press, Cambridge, 1982), p. 156.

The free energy change for the dissolution of NaCl, ΔG_{solv}, is negative, hence we would expect NaCl to dissolve to the extent governed by the magnitude of ΔG_{solv}. A substance with a more negative ΔG_{solv} would be more soluble and one with a less negative or even positive ΔG_{solv} would be less soluble. We will come back to free energies and their relationship to the extent of a reaction in Session 4.

Table 3D-3 Hydration Enthalpies (ΔH_{hyd}) of Common Ions for the Reaction: ION (g) ----> ION (aq)

Ion	ΔH_{hyd} (kJ/mol)	Ion	ΔH_{hyd} (kJ/mol)
Cations, M$^+$		Zn^{+2}	-2,047
H$^+$	-1,091	Cd^{+2}	-1,809
Li$^+$	-520	Hg^{+2}	-1,829
Na$^+$	-406	Sn^{+2}	-1,554
K$^+$	-320	Pb^{+2}	-1,485
Rb$^+$	-296	**Cations, M^{+3}**	
Cs$^+$	-264	Al^{+3}	-4,680
Ag$^+$	-468	Sc^{+3}	-3,930
Tl$^+$	-328	Y^{+3}	-4,105
Cations, M^{+2}		Ga^{+3}	-4,701
Be^{+2}	-2,484	In^{+3}	-4,118
Mg^{+2}	-1,926	Tl^{+3}	-4,108
Ca^{+2}	-1,579	Cr^{+3}	-4,563
Sr^{+2}	-1,446	Fe^{+3}	-4,429
Ba^{+2}	-1,309	Co^{+3}	-4,653
Cr^{+2}	-1,908	**Anions, X$^-$**	
Mn^{+2}	-1,851	F$^-$	-524
Fe^{+2}	-1,950	Cl$^-$	-378
Co^{+2}	-2,010	Br$^-$	-348
Ni^{+2}	-2,096	I$^-$	-308
Cu^{+2}	-2,099		

Source: W. E. Dasent, *Inorganic Energetics*, 2nd edition (Cambridge University Press, Cambridge, 1982), p. 152.

Developing Ideas

Go back to your work on Exploration 3A and take a closer look at the Set 1 compounds: LiCl, NaOH, FeCl$_3$, KI, MgO, ZnO, NaF, CuS, CaCl$_2$, .

1. Group the compounds in two different ways:
   - by the charge on the cation
   - by the charge on the anion

   How does the solubility of the compounds correlate to the groupings you have just made? Interpret your results in terms of Coulomb's law.

2. The expanded crystal lattices of two different ionic solids, CsCl and NaCl, are shown below. For both compounds, count the number of cations that are immediately adjacent to the central chloride anion. List at least one way in which the structure of the crystal lattice could affect the lattice enthalpy, ΔH_L.

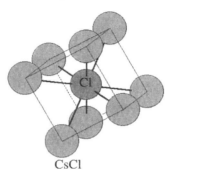

CsCl

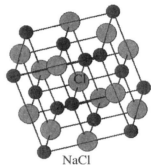

NaCl

3. The ionic radii of several ions are given in Table 3D-4. Assuming the ions in the crystal lattice are touching each other, calculate r, the interionic distance, for the ions in the crystal lattices of TiO$_2$, NaCl, NaF, MgO, and CaO. Use Coulomb's law to predict the effect a small r will have on the lattice enthalpy, ΔH_L, of an ionic substance.

Table 3D-4 Radii of Common Ions

Ion	Radius (Å)	Ion	Radius (Å)
Li$^+$	0.90	Ti^{+4}	0.75
Na$^+$	1.16	F$^-$	1.17
K$^+$	1.52	Cl$^-$	1.67
Mg^{+2}	0.86	Br$^-$	1.82
Ca^{+2}	1.14	O^{-2}	1.26
Fe^{+3}	0.78	S^{-2}	1.70
Fe^{+2}	0.92	OH$^-$	1.19
Zn^{+2}	0.74	CO$_3^{-2}$	1.64
Cu^{+2}	0.71	HCO$_3^-$	1.42
Al^{+3}	0.68	NO$_3^-$	1.65
Si^{+4}	0.50	SO$_4^{-2}$	2.44

Source: *Inorganic Chemistry*, J. E. Huheey, E. A. Keiter, and R. L. Keiter, 4th edition (Harper Collins College Publishers, 1993), pp. 117-119 and p. 118.

4. Based on your work in problems 1-3, list three factors that affect the energy of attraction between ions in a crystal lattice.

5. Table 3D-3 gives the enthalpies of hydration, ΔH_{hyd}, for a variety of ions.

   a. How does ΔH_{hyd} change with respect to the following parameters:
   - The charge on the ion
   - The size of the ion

   b. Explain the trends in terms of Coulomb's law.

6. Table 3D-2 gives the entropies of hydration, ΔS_{hyd}, for a variety of ions.

   a. How does ΔS_{hyd} change with respect to the size of the ion?

   b. The ΔS_{hyd} for F^- is very negative relative to the other comparable anions Cl^-, Br^-, and I^-. Speculate on why this is true.

Working with the Ideas

The following problems will help you think about the concepts further.

7. The ions in a crystal lattice experience repulsive as well as attractive forces. Overall, these repulsive forces are weaker than the attractive forces or the lattice would not stay together. Look at the expanded picture of NaCl in Problem 2 and label all cations and anions in the structure with "+" and "-". Give an example of where repulsive forces would be experienced and explain why these forces are weaker than the primary attractive forces between ions.

8. The lattice enthalpy, ΔH_L, of the silver halides decreases in the order:

   $$AgF > AgCl > AgBr > AgI$$

   However, the solubility of these compounds decreases in the order:

   $$AgF > AgCl > AgBr > AgI$$

   Explain how this can be true.

9. Many salts produce heat when they dissolve in water, i.e., ΔH_{solv} is negative. Is it possible for a substance to dissolve, yet get cold on dissolution? Explain.

Chemical Principles

What chemistry experience have you gained?

Thermodynamics of Dissolution Reactions

You should be able to:

- Describe dissolution in terms of enthalpy changes (bond making and bond breaking, recognizing that bond breaking requires energy and bond making releases energy) (**Exploration 3C**).

- Recognize the difference between the lattice free energy (ΔG_L), the free energy of solvation (ΔG_{solv}), and the free energy of hydration (ΔG_{hyd}) and show how they are interrelated (**Exploration 3D**).

- Describe the factors that affect the magnitude of ΔH_L. These include: (1) the *size* of the ions in the lattice, with smaller ions typically having more negative lattice energies because the attractive energy between ions is stronger when they are closer, (2) the *charge* on the ions in the lattice, with more highly charged ions typically having more negative lattice energies because the attractive energy between ions of greater charge is stronger than between ions of lesser charge, and (3) the arrangement (or packing) of ions in the lattice (**Exploration 3D**).

- Recognize that dissolution will not occur unless ΔG_{solv} is negative. The more negative ΔG_{solv}, the greater the solubility of the compound. The relative contributions of ΔH_{solv} and ΔS_{solv} are different for different compounds and often oppose each other. Thus, a careful analysis of the thermodynamics of dissolution is necessary to explain observed solubilities (**Exploration 3D**).

You may wish to look up these topics in your introductory chemistry textbook for additional explanations and examples. They will most likely be discussed in chapters on solution chemistry and thermodynamics and on intermolecular forces.

Thinking Skills

What general skills are you building for your resume?

As you worked through Session 3, you have also been developing some general problem solving and scientific thinking skills that are valued by employers in a wide range of professions and in academia. Here is a list of the skills that you have been building for your resume.

Analogical and Deductive Reasoning Skills

You should be able to:

- Gather and organize data to look for trends and differences (**Exploration 3C and 3D**).
- Apply a set of principles learned in one setting to make predictions in a different setting (**Explorations 3B and 3C**).
- Use existing data to draw analogies to a new situation (**Exploration 3C and 3D**).

Data analysis skills

You are able to:

- Use graphical methods (bar plots) to compare the relative magnitude of different parameters affecting a system (**Exploration 3D**).
- Use tables of data to look for trends and make predictions (**Exploration 3D**).